Excel

YEARS 8 to 10

Step-by-Step Algebra 2 Workbook

ESSENTIAL skills

Get the Results You Want!

PASCAL PRESS

Lyn Baker

Reprinted 2001, 2002, 2004, 2006, 2007, 2008, 2010, 2012, 2013, 2014, 2015, 2016, 2020 (twice), 2021, 2022, 2023

ISBN 978 1 74020 041 7

Pascal Press
PO Box 250
Glebe NSW 2037
(02) 9198 1748
www.pascalpress.com.au

Publisher: Vivienne Joannou
Edited by Ken Tate
Typeset by Big Picture Communications
Cover by Dizign Pty Ltd
Printed by Vivar Printing/Green Giant Press

TABLE OF CONTENTS

INTRODUCTION

This book is the second in a series of three on algebra and builds on the knowledge gained from the first book. A review of the important basic concepts is included. While it would be an advantage to have worked through Book 1, students with some knowledge of basic algebra should be able to understand and benefit from this book.

Each section is devoted to one aspect of the course and consists of four pages. The first page explains the content and gives plenty of examples, which should be carefully read and understood. This is followed by a question page, which should be completed using the examples as a guide. The next two pages provide fully worked solutions to every question, with further explanations and hints.

At the end of the book you will find a glossary of common words used in algebra and their meanings.

2 REVIEW OF BASIC OPERATIONS

Like terms are terms whose pronumeral parts are *exactly* the same.
Only like terms can be added together or subtracted. Unlike terms cannot be simplified.

The operation sign belonging to a term is the one immediately before the term.

Example:

Find:

(a) $2a + 3a$
(b) $3x + 4y$
(c) $-6k + 2k$
(d) $2ab - 5ab$
(e) $7x^2 + 2x^2 - 9$
(f) $e^5 + e^5$
(g) $8t - 12t + 3t$

Solution:

(a) $2a + 3a = \mathbf{5a}$
(b) $3x + 4y = \mathbf{3x + 4y}$
(c) $-6k + 2k = \mathbf{-4k}$
(d) $2ab - 5ab = \mathbf{-3ab}$
(e) $7x^2 + 2x^2 - 9 = \mathbf{9x^2 - 9}$
(f) $e^5 + e^5 = \mathbf{2e^5}$
(g) $8t - 12t + 3t = \mathbf{-t}$ ($-t$ means $-1t$; we don't need to write the 1.)

The order in which we write the answer doesn't matter as long as the correct signs are attached to each term: ($-3x + 3y$ or $3y - 3x$; $-$ before $3x$, $+$ before $3y$)

Example:

Find:

(a) $8g + 5 - 3g - 11$
(b) $2x + 3y - 5x$
(c) $6b^2 - 3b^3 + 4b^2 - 5b^3$
(d) $4m^3 + 2m^2 - 5m^3 - m^2 + m$
(e) $n^4 + 4 - 3n^3 + 2n - 3n^4 + 2n^2 - 5n - 3$

Solution:

(a) $8g + 5 - 3g - 11 = \mathbf{5g - 6}$
(b) $2x + 3y - 5x = \mathbf{-3x + 3y}$
(c) $6b^2 - 3b^3 + 4b^2 - 5b^3 = \mathbf{10b^2 - 8b^3}$
(d) $4m^3 + 2m^2 - 5m^3 - m^2 + m = \mathbf{-m^3 + m^2 + m}$
(e) $n^4 + 4 - 3n^3 + 2n - 3n^4 + 2n^2 - 5n - 3$
$= \mathbf{-2n^4 - 3n^3 + 2n^2 - 3n + 1}$

When multiplying pronumerals we don't need to write the multiplication sign.
$7 \times a = 7a$ $-3 \times e = -3e$ $-1 \times x = -x$ $5k$ means $5 \times k$

Example:

Find:

(a) $2 \times 3x$
(b) $3 \times -8k$
(c) $-4 \times 5b$
(d) $-x \times -y$
(e) $k \times -k$
(f) $3 \times 2a \times -4b$
(g) $2x \times 3x \times 4x$
(h) $8y \div 8$
(i) $6p \div p$
(j) $-9k \div 3$
(k) $10ab \div -5b$
(l) $-12stu \div -3u$

Solution:

(a) $2 \times 3x = \mathbf{6x}$
(b) $3 \times -8k = \mathbf{-24k}$
(c) $-4 \times 5b = \mathbf{-20b}$
(d) $-x \times -y = \mathbf{xy}$
(e) $k \times -k = \mathbf{-k^2}$
(f) $3 \times 2a \times -4b = \mathbf{-24ab}$
(g) $2x \times 3x \times 4x = \mathbf{24x^3}$
(h) $8y \div 8 = \mathbf{y}$
(i) $6p \div p = \mathbf{6}$
(j) $-9k \div 3 = \mathbf{-3k}$
(k) $10ab \div -5b = \mathbf{-2a}$
(l) $-12stu \div -3u = \mathbf{4st}$

1. **Add or subtract:**

(a) $6p + 3p$ = ______
(b) $9t + t$ = ______
(c) $-3e + 8e$ = ______
(d) $-5y + 2y$ = ______
(e) $6x^2 - 3x^2$ = ______
(f) $7g^3 - 6g^3$ = ______
(g) $5s - 7s$ = ______
(h) $2q - 9q$ = ______
(i) $-a + a$ = ______
(j) $-3c - 2c$ = ______
(k) $-7q^2 + q^2$ = ______
(l) $-4y + 5y$ = ______
(m) $-b^3 + 6b^3$ = ______
(n) $-4h^2 - 6h^2$ = ______
(o) $4ab - 5ab$ = ______
(p) $-k - k$ = ______

2. **Simplify, where possible:**

(a) $4e + 3e$ = ______
(b) $6x + 3y$ = ______
(c) $9k - k$ = ______
(d) $-2n + 4n$ = ______
(e) $-5j - 3j$ = ______
(f) $4m^2 + 3m$ = ______
(g) $4p - 7q$ = ______
(h) $-3t^2 + 3t^2$ = ______
(i) $-2a - 6b$ = ______
(j) $-3r - 6r$ = ______
(k) $-5m^3 + 6m^2$ = ______
(l) $-4t + 9$ = ______

3. **Simplify by collecting like terms:**

(a) $9y + 3z + 2y$ = ______
(b) $2g + 4h - 5g + 7h$ = ______
(c) $16 - 3x + 7x - 4$ = ______
(d) $3t^2 - 5t - 8t$ = ______
(e) $5b + 3c - 2b - 7c$ = ______
(f) $6a^2 - 2a^3 + 7a^3$ = ______
(g) $-3p + 7q + p - 2q$ = ______
(h) $-7n + 4 - 3n - 9$ = ______
(i) $4d - 3d^2 - 6d^2 + d^2$ = ______
(j) $k - 1 + 2k - 5 - 4$ = ______
(k) $5m - 8m - 2m + 6n$ = ______
(l) $-3r^2 - r^2 + 2r - 7r^3$ = ______
(m) $e^2 + e^3 - e^4 + e^3 - e^2$ = ______
(n) $n^2 - 2n - 3n^2 + 8n - 6n$ = ______

4. **Find:**

(a) $4 \times 2n$ = ______
(b) $7 \times -3m$ = ______
(c) $-5 \times 4b^2$ = ______
(d) $-2 \times -3g$ = ______
(e) $-p \times q$ = ______
(f) $2e \times -6f$ = ______
(g) $-8t \times 9$ = ______
(h) $-a \times -b$ = ______
(i) $3c \times -5c$ = ______
(j) $-y \times 7yz$ = ______
(k) $u \times -4v$ = ______
(l) $-2a \times -3a \times -4a$ = ______
(m) $4q \div 4$ = ______
(n) $-5t \div -5$ = ______
(o) $3j \div -3$ = ______
(p) $-8n \div -4$ = ______
(q) $7h \div h$ = ______
(r) $-9w \div w$ = ______
(s) $5c \div -c$ = ______
(t) $10m \div -2m$ = ______
(u) $-14ab \div 7a$ = ______
(v) $6pq \div -2q$ = ______
(w) $-9xy \div -3y$ = ______
(x) $8cde \div -4ce$ = ______

REVIEW OF BASIC OPERATIONS Worked Solutions

1. Add or subtract:

(a) $6p + 3p = \mathbf{9p}$

6 lots of p plus 3 lots of p = 9 lots of p.

(b) $9t + t = \mathbf{10t}$

$t = 1t$, $9t + 1t = 10t$

(c) $-3e + 8e = \mathbf{5e}$

(d) $-5y + 2y = \mathbf{-3y}$

(e) $6x^2 - 3x^2 = \mathbf{3x^2}$

6 lots of x^2 – 3 lots of x^2 = 3 lots of x^2.

(f) $7g^3 - 6g^3 = \mathbf{g^3}$

7 lots of g^3 – 6 lots of g^3 = 1 lot of g^3.
We don't need to write the 1.

(g) $5s - 7s = \mathbf{-2s}$

(h) $2q - 9q = \mathbf{-7q}$

(i) $-a + a = \mathbf{0}$

0 lots of a = 0.

(j) $-3c - 2c = \mathbf{-5c}$

(k) $-7q^2 + q^2 = \mathbf{-6q^2}$

(l) $-4y + 5y = \mathbf{y}$

(m) $-b^3 + 6b^3 = \mathbf{5b^3}$

(n) $-4h^2 - 6h^2 = \mathbf{-10h^2}$

(o) $4ab - 5ab = \mathbf{-ab}$

4 lots of ab – 5 lots of ab = –1 lots of ab.
$-ab$ means $-1ab$.

(p) $-k - k = \mathbf{-2k}$

2. Simplify, where possible:

(a) $4e + 3e = \mathbf{7e}$

(b) $6x + 3y = \mathbf{6x + 3y}$

$6x$ and $3y$ are unlike terms.
This can't be simplified.

(c) $9k - k = \mathbf{8k}$

(d) $-2n + 4n = \mathbf{2n}$

(e) $-5j - 3j = \mathbf{-8j}$

(f) $4m^2 + 3m = \mathbf{4m^2 + 3m}$

Although $4m^2$ and $3m$ use the same pronumeral, their pronumeral parts are not exactly the same.

(g) $4p - 7q = \mathbf{4p - 7q}$

(h) $-3t^2 + 3t^2 = \mathbf{0}$

(i) $-2a - 6b = \mathbf{-2a - 6b}$

(j) $-3r - 6r = \mathbf{-9r}$

(k) $-5m^3 + 6m^2 = \mathbf{-5m^3 + 6m^2}$

They are unlike terms.

(l) $-4t + 9 = \mathbf{-4t + 9}$

Unlike terms.

3. Simplify by collecting like terms:

(a) $9y + 3z + 2y = \mathbf{11y + 3z}$

or $3z + 11y$

(b) $2g + 4h - 5g + 7h = \mathbf{-3g + 11h}$

(c) $16 - 3x + 7x - 4 = \mathbf{12 + 4x}$ ☞ or $4x + 12$

(d) $3t^2 - 5t - 8t = \mathbf{3t^2 - 13t}$

(e) $5b + 3c - 2b - 7c = \mathbf{3b - 4c}$ ☞ $5b + 3c - 2b - 7c$
$5b - 2b = 3b$, $3c - 7c = -4c$

(f) $6a^2 - 2a^3 + 7a^3 = \mathbf{6a^2 + 5a^3}$ ☞ $6a^2 - 2a^3 + 7a^3$

(g) $-3p + 7q + p - 2q = \mathbf{-2p + 5q}$ ☞ $-3p + 7q + p - 2q$

(h) $-7n + 4 - 3n - 9 = \mathbf{-10n - 5}$ ☞ $-7n + 4 - 3n - 9$

(i) $4d - 3d^2 - 6d^2 + d^2 = \mathbf{4d - 8d^2}$ ☞ or $-8d^2 + 4d$

(j) $k - 1 + 2k - 5 - 4 = \mathbf{3k - 10}$

(k) $5m - 8m - 2m + 6n = \mathbf{-5m + 6n}$

(l) $-3r^2 - r^2 + 2r - 7r^3 = \mathbf{-7r^3 - 4r^2 + 2r}$

(m) $e^2 + e^3 - e^4 + e^3 - e^2 = \mathbf{2e^3 - e^4}$

(n) $n^2 - 2n - 3n^2 + 8n - 6n = \mathbf{-2n^2}$

4. Find:

(a) $4 \times 2n = \mathbf{8n}$ ☞ $2n + 2n + 2n + 2n = 8n$ 4 lots of $2n = 8n$

(b) $7 \times -3m = \mathbf{-21m}$

(c) $-5 \times 4b^2 = \mathbf{-20b^2}$

(d) $-2 \times -3g = \mathbf{6g}$

(e) $-p \times q = \mathbf{-pq}$

(f) $2e \times -6f = \mathbf{-12ef}$

(g) $-8t \times 9 = \mathbf{-72t}$

(h) $-a \times -b = \mathbf{ab}$

(i) $3c \times -5c = \mathbf{-15c^2}$ ☞ c^2 means $c \times c$

(j) $-y \times 7yz = \mathbf{-7y^2z}$

(k) $u \times -4v = \mathbf{-4uv}$

(l) $-2a \times -3a \times -4a = \mathbf{-24a^3}$

(m) $4q \div 4 = \mathbf{q}$ ☞ $4q$ means $4 \times q$ so $4q \div 4 = q$

(n) $-5t \div -5 = \mathbf{t}$

(o) $3j \div -3 = \mathbf{-j}$

(p) $-8n \div -4 = \mathbf{2n}$

(q) $7h \div h = \mathbf{7}$

(r) $-9w \div w = \mathbf{-9}$

(s) $5c \div -c = \mathbf{-5}$

(t) $10m \div -2m = \mathbf{-5}$

(u) $-14ab \div 7a = \mathbf{-2b}$

(v) $6pq \div -2q = \mathbf{-3p}$

(w) $-9xy \div -3y = \mathbf{3x}$

(x) $8cde \div -4ce = \mathbf{-2d}$

6 REVIEW OF THE INDEX RULES AND MIXED OPERATIONS

When multiplying, if the base is the same, add the indices. $a^x \times a^y = a^{x+y}$
When dividing, if the base is the same, subtract the indices. $a^x \div a^y = a^{x-y}$
When raising a power to a power, multiply the indices. $(a^x)^y = a^{xy}$
Anything raised to the zero power equals one. $a^0 = 1$

Example:

Simplify:

(a) $x^3 \times x^7$
(b) $7t^2 \times 3t^5$
(c) $-6e^4 \times -2e^3$
(d) $9d^9 \div 3d^3$
(e) $15h^7 \div -3h^6$
(f) $-2t^2 \times 3t^3 \times -4t^4$
(g) $\frac{xy^4}{xy^2}$
(h) $\frac{2k^2}{4k}$
(i) $\frac{n^2}{n^3}$

Solution:

(a) $x^3 \times x^7 = \mathbf{x^{10}}$
(b) $7t^2 \times 3t^5 = \mathbf{21t^7}$
(c) $-6e^4 \times -2e^3 = \mathbf{12e^7}$
(d) $9d^9 \div 3d^3 = \mathbf{3d^6}$
(e) $15h^7 \div -3h^6 = \mathbf{-5h}$
(f) $-2t^2 \times 3t^3 \times -4t^4 = \mathbf{24t^9}$
(g) $\frac{xy^4}{xy^2} = \mathbf{y^2}$
(h) $\frac{2k^2}{4k} = \mathbf{\frac{k}{2}}$
(i) $\frac{n^2}{n^3} = \mathbf{\frac{1}{n}}$

> In questions like $(3x^4)^2$, any numbers inside the brackets must be raised to the power.
> In $(3x^4)^2$, the 3 must be squared. $(3x^4)^2 = 9x^8$.

Example:

Find:

(a) $(-2k^5)^2$
(b) $(-2g^2)^3$
(c) $-3(n^4)^4$
(d) m^0
(e) $-2v^0$
(f) $(a^2b^3)^4$
(g) $(5b)^0$

Solution:

(a) $(-2k^5)^2 = \mathbf{4k^{10}}$
(b) $(-2g^2)^3 = \mathbf{-8g^6}$
(c) $-3(n^4)^4 = \mathbf{-3n^{16}}$
(d) $m^0 = \mathbf{1}$
(e) $-2v^0 = \mathbf{-2}$
(f) $(a^2b^3)^4 = \mathbf{a^8b^{12}}$
(g) $(5b)^0 = \mathbf{1}$

We must take care to perform operations in the correct order.

Example:

(a) $-6m^5 \times (2m)^2$
(b) $4h^0 - (5u)^0$
(c) $12z^{12} \div 4z^4 \times 2z^2$
(d) $(4d^2)^3 \div 8d^2$
(e) $7e^2 \times 3e^2 + 5e \times 2e^3$
(f) $\left(\frac{a^8}{a^2}\right)^2$

Solution:

(a) $-6m^5 \times (2m)^2 = \mathbf{-6m^5 \times 4m^2}$
$= \mathbf{-24m^7}$

(b) $4h^0 - (5u)^0 = \mathbf{4 - 1}$
$= \mathbf{3}$

(c) $12z^{12} \div 4z^4 \times 2z^2 = \mathbf{3z^8 \times 2z^2}$
$= \mathbf{6z^{10}}$

(d) $(4d^2)^3 \div 8d^2 = \mathbf{64d^6 \div 8d^2}$
$= \mathbf{8d^4}$

(e) $7e^2 \times 3e^2 + 5e \times 2e^3 = \mathbf{21e^4 + 10e^4}$
$= \mathbf{31e^4}$

(f) $\left(\frac{a^8}{a^2}\right)^2 = \mathbf{(a^6)^2}$
$= \mathbf{a^{12}}$

1. **Simplify:**

(a) $3a^4 \times 2a^3$ = ____________

(b) $-4x^2 \times -5x$ = ____________

(c) $-3p^5 \times 7p^3$ = ____________

(d) $2y^4 \times 2y^6$ = ____________

(e) $2x^2y^3 \times -5xy^3$ = ____________

(f) $-3ab^2 \times -4a^2b^2$ = ____________

(g) $12d^{12} \div 2d^2$ = ____________

(h) $4h^8 \div -2h^4$ = ____________

(i) $-6k^9 \div -3k^4$ = ____________

(j) $-10y^{10} \div 5y$ = ____________

(k) $p^6q^8 \div p^2q^7$ = ____________

(l) $16t^4u^4 \div 4tu^2$ = ____________

(m) h^0 = ____________

(n) $7n^0$ = ____________

(o) $-4r^0$ = ____________

(p) $(-8w)^0$ = ____________

(q) $(a^3)^3$ = ____________

(r) $(2e)^5$ = ____________

(s) $(-3t^2)^2$ = ____________

(t) $-4(g^7)^3$ = ____________

(u) $(-2a^4)^3$ = ____________

(v) $(ab^5)^4$ = ____________

(w) $-(m^3n^2)^4$ = ____________

2. **Simplify:**

(a) $(a^2)^4 \times (a^3)^2$ = ____________
= ____________

(b) $(6e^5)^2 \div -12e^4$ = ____________
= ____________

(c) $-6b^6 \times 4b^4 \div 8b^8$ = ____________
= ____________

(d) $-(3u^3)^4 \div 9u^7$ = ____________
= ____________

(e) $(-4d)^2 \div -4d^2$ = ____________
= ____________

(f) $-12h^7 \div -2h^3 \times 4h$ = ____________
= ____________

(g) $-4b^5 \times 7b^3 + -9b^3 \div -9b$ = ____________

(h) $3x^2y^4 \times -5xy^3 \div 15x^3y^7$ = ____________
= ____________

(i) $(-2a^3)^2 \div (2a^3)^2$ = ____________
= ____________

(j) $(-6x^5)^3 \div 2x^9 \div 3x^4$ = ____________
= ____________
= ____________

(k) $-6a^4b^3 \times -2ab^2 \div 4(ab)^2$ = ____________
= ____________

(l) $-3(t^4)^3 \times -2(t^3)^5 + (2t^3)^3$ = ____________
= ____________

(m) $4c^5 \times (-2c)^4 \div (-c^3)^3$ = ____________
= ____________
= ____________

(n) $3(g^2)^2 \times (5g)^0 + 5g^0 \times (3g^2)^2$ = ____________
= ____________
= ____________

(o) $(2p^4)^2 \times (-2p^3)^4 - 6p^{11} \times -3p^7$ = ____________
= ____________
= ____________

8 INDEX RULES AND MIXED OPERATIONS Worked Solutions

1. Simplify:

(a) $3a^4 \times 2a^3 = \mathbf{6a^7}$ — Multiply the numbers, add the indices.

(b) $-4x^2 \times -5x = \mathbf{20x^3}$ — x means x^1; $x^2 \times x = x^{2+1}$

(c) $-3p^5 \times 7p^3 = \mathbf{-21p^8}$

(d) $2y^4 \times 2y^6 = \mathbf{4y^{10}}$

(e) $2x^2y^3 \times -5xy^3 = \mathbf{-10x^3y^6}$ — Add the indices, separately for each pronumeral.

(f) $-3ab^2 \times -4a^2b^2 = \mathbf{12a^3b^4}$

(g) $12d^{12} \div 2d^2 = \mathbf{6d^{10}}$ — Divide the numbers, subtract the indices.

(h) $4h^8 \div -2h^4 = \mathbf{-2h^4}$

(i) $-6k^9 \div -3k^4 = \mathbf{2k^5}$

(j) $-10y^{10} \div 5y = \mathbf{-2y^9}$

(k) $p^6q^8 \div p^2q^7 = \mathbf{p^4q}$

(l) $16t^4u^4 \div 4tu^2 = \mathbf{4t^3u^2}$

(m) $h^0 = \mathbf{1}$ — Anything raised to the zero power equals 1.

(n) $7n^0 = \mathbf{7}$ — $7n^0 = 7 \times n^0 = 7 \times 1$

(o) $-4r^0 = \mathbf{-4}$ — $-4r^0 = -4 \times r^0 = -4 \times 1$

(p) $(-8w)^0 = \mathbf{1}$ — Anything raised to the zero power equals 1.

(q) $(a^3)^3 = \mathbf{a^9}$ — Multiply the indices.

(r) $(2e)^5 = \mathbf{32e^5}$ — 2 and e must both be raised to the fifth power. $2^5 = 32$

(s) $(-3t^2)^2 = \mathbf{9t^4}$ — $(-3)^2 = -3 \times -3 = 9$

(t) $-4(g^7)^3 = \mathbf{-4g^{21}}$ — -4 is not in the bracket, so it is not cubed.

(u) $(-2a^4)^3 = \mathbf{-8a^{12}}$

(v) $(ab^5)^4 = \mathbf{a^4b^{20}}$ — $(a^1b^5)^4 = a^{1\times4} \times b^{5\times4}$

(w) $-(m^3n^2)^4 = \mathbf{-m^{12}n^8}$ — $-(m^3n^2)^4 = -1(m^3n^2)^4$. -1 is not in the bracket, so it is not raised to the fourth power.

2. Simplify:

(a) $(a^2)^4 \times (a^3)^2 = \mathbf{a^8 \times a^6}$
$= \mathbf{a^{14}}$ — Raising a power to a power, multiply the indices. When multiplying, add the indices.

(b) $(6e^5)^2 \div -12e^4 = \mathbf{36e^{10} \div -12e^4}$

$= \mathbf{-3e^6}$

Raising a power to a power, multiply the indices. When dividing, subtract the indices.

(c) $-6b^6 \times 4b^4 \div 8b^8 = \mathbf{-24b^{10} \div 8b^8}$

$= \mathbf{-3b^2}$

(d) $-(3u^3)^4 \div 9u^7 = \mathbf{-81u^{12} \div 9u^7}$

$= \mathbf{-9u^5}$

(e) $(-4d)^2 \div -4d^2 = \mathbf{16d^2 \div -4d^2}$

$= \mathbf{-4}$

$d^2 \div d^2 = d^0 = 1, \quad -4 \times 1 = -4$

(f) $-12h^7 \div -2h^3 \times 4h = \mathbf{6h^4 \times 4h}$

$= \mathbf{24h^5}$

(g) $-4b^5 \times 7b^3 + -9b^3 \div -9b = \mathbf{-28b^8 + b^2}$

Unlike terms cannot be simplified.

(h) $3x^2y^4 \times -5xy^3 \div 15x^3y^7 = \mathbf{-15x^3y^7 \div 15x^3y^7}$

$= \mathbf{-1}$

(i) $(-2a^3)^2 \div (2a^3)^2 = \mathbf{4a^6 \div 4a^6}$

$= \mathbf{1}$

(j) $(-6x^5)^3 \div 2x^9 \div 3x^4 = \mathbf{-216x^{15} \div 2x^9 \div 3x^4}$

$= \mathbf{-108x^6 \div 3x^4}$

$= \mathbf{-36x^2}$

Remember to work from left to right.

(k) $-6a^4b^3 \times -2ab^2 \div 4(ab)^2 = \mathbf{12a^5b^5 \div 4a^2b^2}$

$= \mathbf{3a^3b^3}$

(l) $-3(t^4)^3 \times -2(t^3)^5 + (2t^3)^3 = \mathbf{-3t^{12} \times -2t^{15} + 8t^9}$

$= \mathbf{6t^{27} + 8t^9}$

Unlike terms can't be simplified.

(m) $4c^5 \times (-2c)^4 \div (-c^3)^3 = \mathbf{4c^5 \times 16c^4 \div -c^9}$

$(-c^3)^3 = (-1c^3)^3, \quad (-1)^3 = -1 \quad \text{so} \quad (-c^3)^3 = -c^9$

$= \mathbf{64c^9 \div -c^9}$

$= \mathbf{-64}$

$c^0 = 1$

(n) $3(g^2)^2 \times (5g)^0 + 5g^0 \times (3g^2)^2 = \mathbf{3g^4 \times 1 + 5 \times 9g^4}$

$= \mathbf{3g^4 + 45g^4}$

$= \mathbf{48g^4}$

(o) $(2p^4)^2 \times (-2p^3)^4 - 6p^{11} \times -3p^7 = \mathbf{4p^8 \times 16p^{12} + 18p^{18}}$

$= \mathbf{64p^{20} + 18p^{18}}$

10 REMOVING BRACKETS

We know that $2(x + 8) = 2 \times x + 2 \times 8$
$= 2x + 16$

We multiply every term inside the bracket by the number, or term, outside the bracket.

Example:

Expand:

(a) $3(4p + 5q)$ **(b)** $9(8x^2 + 4x - 1)$

Solution:

(a) $3(4p + 5q) = \mathbf{12p + 15q}$ (b) $9(8x^2 + 4x - 1) = \mathbf{72x^2 + 36x - 9}$

We use the same rule if the number, or term, outside the brackets is negative.

$-3(e + 4) = -3 \times e + (-3 \times 4)$
$= -3e + (-12)$
$= -3e - 12$

$-7(3x - 2) = -7 \times 3x - (-7 \times 2)$
$= -21x - (-14)$
$= -21x + 14$

We don't need to write out all those steps. Just remember:
A negative outside a bracket will change any signs inside the bracket.

Example:

(a) $-2(x + y)$
(b) $-a(2a - 7)$
(c) $-5(3p - 4q)$
(d) $-3(9 + z)$
(e) $-3a(b - c)$
(f) $-4t(2t - 5)$

Solution:

(a) $-2(x + y) = \mathbf{-2x - 2y}$
(b) $-a(2a - 7) = \mathbf{-2a^2 + 7a}$
(c) $-5(3p - 4q) = \mathbf{-15p + 20q}$
(d) $-3(9 + z) = \mathbf{-27 - 3z}$
(e) $-3a(b - c) = \mathbf{-3ab + 3ac}$
(f) $-4t(2t - 5) = \mathbf{-8t^2 + 20t}$

Every sign inside the bracket will be changed if we multiply by a negative number.

Example:

Expand:

(a) $-8(a + b - c)$ **(b)** $-5(5x - 2y - 4z + 9)$

Solution:

(a) $-8(a + b - c) = \mathbf{-8a - 8b + 8c}$ (b) $-5(5x - 2y - 4z + 9) = \mathbf{-25x + 10y + 20z - 45}$

$-(t + 4)$ means $-1(t + 4)$ (We don't need to write the 1.)
So $-(t + 4) = -t - 4$ (Multiplying by -1.)

Example:

Find:

(a) $-(3e - 7f)$
(b) $-(8 + 5y)$
(c) $-(k - 4m)$
(d) $-(x - y + z)$
(e) $-(-2 - 5j)$

Solution:

(a) $-(3e - 7f) = \mathbf{-3e + 7f}$
(b) $-(8 + 5y) = \mathbf{-8 - 5y}$
(c) $-(k - 4m) = \mathbf{-k + 4m}$
(d) $-(x - y + z) = \mathbf{-x + y - z}$
(e) $-(-2 - 5j) = \mathbf{2 + 5j}$

1. **Expand:**

(a) $3(x + 5)$ = ____________ (c) $t(t - 6)$ = ____________ (e) $5(2z + 1)$ = ____________

(b) $4(2a + 3b)$ = ____________ (d) $2e(e + 7f)$ = ____________ (f) $6a(b - c)$ = ____________

2. **Expand:**

(a) $-2(4a + 7)$ = ____________ (e) $-7(2a + 1)$ = ____________ (i) $-5a(b + 7c)$ = ____________

(b) $-5(2e - 3)$ = ____________ (f) $-9(1 - y)$ = ____________ (j) $-4c(c - 8)$ = ____________

(c) $-4(5k + 9m)$ = ____________ (g) $-h(h + 8)$ = ____________ (k) $-3p(p - q)$ = ____________

(d) $-6(x - 1)$ = ____________ (h) $-m(m - n)$ = ____________ (l) $-5m^2(n + 1)$ = ____________

3. **Expand:**

(a) $-(a + 4)$ = ____________ (c) $-(2t + 3v)$ = ____________ (e) $-(-7 - y)$ = ____________

(b) $-(3g - 5)$ = ____________ (d) $-(p + q - r)$ = ____________ (f) $-(x^2 - 6)$ = ____________

4. **Expand:**

(a) $-3d(d - 8)$ = ____________ (e) $-a(a - b)$ = ____________ (i) $-2y(y^2 - 3y)$ = ____________

(b) $k(t + u)$ = ____________ (f) $-9t(t + 1)$ = ____________ (j) $-b^2(a^2 - b^2)$ = ____________

(c) $-(z + 3)$ = ____________ (g) $a^2(a + 5)$ = ____________ (k) $-(4 - w)$ = ____________

(d) $5y(y + 2)$ = ____________ (h) $p(p^2 + q^2)$ = ____________ (l) $-7x^2(x^6 + x^4)$ = ____________

5. **Expand:**

(a) $5(a^5 + 4a^4 - 3a^2 + 2a + 1)$

= ____________________

(e) $4p(p^3 + 2p^2 - 5p - 9)$

= ____________________

(i) $-2m^2(m^3 - 4m^2 - 3m + 7n^2)$

= ____________________

(b) $-8(d^3 - 2d^2 + d - 1)$

= ____________________

(f) $d^2(d^8 + 5d^3 - d)$

= ____________________

(j) $-6c^3(4c^4 + 6c^3 + 7c^2 - c + 5)$

= ____________________

(c) $e(e^3 + 5e^2 - e - 1)$

= ____________________

(g) $-3xy(x^2 + xy - y^2)$

= ____________________

(k) $2n^9(n^8 + n^7 - n^5 + 3n^4 - n^2 - 2n + 1)$

= ____________________

(d) $-3x(2x + 8y - 3)$

= ____________________

(h) $-5a(a^2b^2 + 3ab - b^2)$

= ____________________

12 REMOVING BRACKETS Worked Solutions

1. Expand:

   (a) $3(x+5) = \mathbf{3x + 15}$ — $3 \times x + 3 \times 5$

   (b) $4(2a+3b) = \mathbf{8a + 12b}$ — $4 \times 2a + 4 \times 3b$

   (c) $t(t-6) = \mathbf{t^2 - 6t}$ — $t \times t - t \times 6$

   (d) $2e(e+7f) = \mathbf{2e^2 + 14ef}$

   (e) $5(2z+1) = \mathbf{10z + 5}$

   (f) $6a(b-c) = \mathbf{6ab - 6ac}$

2. Expand:

   (a) $-2(4a+7) = \mathbf{-8a - 14}$ — $-2 \times 4a + (-2 \times 7) = -8a + (-14)$

   (b) $-5(2e-3) = \mathbf{-10e + 15}$ — The negative outside the bracket changes the signs inside the bracket.

   (c) $-4(5k+9m) = \mathbf{-20k - 36m}$

   (d) $-6(x-1) = \mathbf{-6x + 6}$

   (e) $-7(2a+1) = \mathbf{-14a - 7}$

   (f) $-9(1-y) = \mathbf{-9 + 9y}$

   (g) $-h(h+8) = \mathbf{-h^2 - 8h}$

   (h) $-m(m-n) = \mathbf{-m^2 + mn}$

   (i) $-5a(b+7c) = \mathbf{-5ab - 35ac}$

   (j) $-4c(c-8) = \mathbf{-4c^2 + 32c}$

   (k) $-3p(p-q) = \mathbf{-3p^2 + 3pq}$

   (l) $-5m^2(n+1) = \mathbf{-5m^2n - 5m^2}$

3. Expand:

   (a) $-(a+4) = \mathbf{-a - 4}$ — This means $-1(a+4)$. $-1 \times a + (-1 \times 4)$

   (b) $-(3g-5) = \mathbf{-3g + 5}$

   (c) $-(2t+3v) = \mathbf{-2t - 3v}$

   (d) $-(p+q-r) = \mathbf{-p - q + r}$ — Every sign inside the bracket must change.

   (e) $-(-7-y) = \mathbf{7 + y}$

   (f) $-(x^2-6) = \mathbf{-x^2 +}$

4. Expand:

(a) $-3d(d-8) = \mathbf{-3d^2 + 24d}$ — $-3d \times d - (-3d \times 8)$

(b) $k(t+u) = \mathbf{kt + ku}$

(c) $-(z+3) = \mathbf{-z - 3}$ — The signs must change.

(d) $5y(y+2) = \mathbf{5y^2 + 10y}$

(e) $-a(a-b) = \mathbf{-a^2 + ab}$

(f) $-9t(t+1) = \mathbf{-9t^2 - 9t}$

(g) $a^2(a+5) = \mathbf{a^3 + 5a^2}$ — $a^2 \times a + a^2 \times 5$

(h) $p(p^2+q^2) = \mathbf{p^3 + pq^2}$

(i) $-2y(y^2-3y) = \mathbf{-2y^3 + 6y^2}$

(j) $-b^2(a^2-b^2) = \mathbf{-a^2b^2 + b^4}$ — $-b^2a^2 + b^4$ is the same thing, but we usually place pronumerals in alphabetical order.

(k) $-(4-w) = \mathbf{-4 + w}$

(l) $-7x^2(x^6+x^4) = \mathbf{-7x^8 - 7x^6}$ — Don't confuse this type of question involving brackets with raising a power to a power. We are multiplying, so we add the indices.

5. Expand:

(a) $5(a^5 + 4a^4 - 3a^2 + 2a + 1)$
$= \mathbf{5a^5 + 20a^4 - 15a^2 + 10a + 5}$ — Every term in the bracket must be multiplied by 5.

(b) $-8(d^3 - 2d^2 + d - 1) = \mathbf{-8d^3 + 16d^2 - 8d + 8}$ — Every sign in the bracket must change.

(c) $e(e^3 + 5e^2 - e - 1) = \mathbf{e^4 + 5e^3 - e^2 - e}$

(d) $-3x(2x + 8y - 3) = \mathbf{-6x^2 - 24xy + 9x}$

(e) $4p(p^3 + 2p^2 - 5p - 9) = \mathbf{4p^4 + 8p^3 - 20p^2 - 36p}$

(f) $d^2(d^8 + 5d^3 - d) = \mathbf{d^{10} + 5d^5 - d^3}$

(g) $-3xy(x^2 + xy - y^2) = \mathbf{-3x^3y - 3x^2y^2 + 3xy^3}$

(h) $-5a(a^2b^2 + 3ab - b^2) = \mathbf{-5a^3b^2 - 15a^2b + 5ab^2}$

(i) $-2m^2(m^3 - 4m^2 - 3m + 7n^2)$
$= \mathbf{-2m^5 + 8m^4 + 6m^3 - 14m^2n^2}$ — To multiply (if the base is the same), add the indices.

(j) $-6c^3(4c^4 + 6c^3 + 7c^2 - c + 5)$
$= \mathbf{-24c^7 - 36c^6 - 42c^5 + 6c^4 - 30c^3}$

(k) $2n^9(n^8 + n^7 - n^5 + 3n^4 - n^2 - 2n + 1)$
$= \mathbf{2n^{17} + 2n^{16} - 2n^{14} + 6n^{13} - 2n^{11} - 4n^{10} + 2n^9}$

14 EXPANDING AND SIMPLIFYING

After we have expanded grouping symbols we must, if possible, simplify the expression by collecting like terms.

Example:

Expand and simplify:

(a) $3(g-8)+5(g+2)$

(b) $7(3d+2)+4(6d-5)$

Solution:

(a) $3(g-8)+5(g+2) = \mathbf{3g - 24 + 5g + 10}$

$= \mathbf{8g - 14}$

(b) $7(3d+2)+4(6d-5) = \mathbf{21d + 14 + 24d - 20}$

$= \mathbf{45d - 6}$

When subtracting a bracket, we must take care to subtract every term in the bracket. Any operation signs in that bracket will change.

$5x - 2(3x + 4) = 5x - 6x - 8$
$= -x - 8$

$2t - (3 - t) = 2t - 3 + t$
$= 3t - 3$

> We need to be extra careful when working with this type of question. It is very easy to make a mistake with the signs!

Example:

Expand and simplify:

(a) $2(3x+4y)-5(3x-4y)$

(b) $8(2p+7)-3(6p+1)$

(c) $9(4k-3)-12(3k+2)$

(d) $6(m-n)-(2m+n)$

(e) $3(4a+7)-(7-a)$

(f) $4(a+b-c)-3(a-b-c)$

Solution:

(a) $2(3x+4y)-5(3x-4y) = \mathbf{6x + 8y - 15x + 20y}$

$= \mathbf{-9x + 28y}$

(b) $8(2p+7)-3(6p+1) = \mathbf{16p + 56 - 18p - 3}$

$= \mathbf{-2p + 53}$

(c) $9(4k-3)-12(3k+2) = \mathbf{36k - 27 - 36k - 24}$

$= \mathbf{-51}$

(d) $6(m-n)-(2m+n) = \mathbf{6m - 6n - 2m - n}$

$= \mathbf{4m - 7n}$

(e) $3(4a+7)-(7-a) = \mathbf{12a + 21 - 7 + a}$

$= \mathbf{13a + 14}$

(f) $4(a+b-c)-3(a-b-c)$

$= \mathbf{4a + 4b - 4c - 3a + 3b + 3c}$

$= \mathbf{a + 7b - c}$

1. **Expand and simplify:**

(a) $4(a - 6) + 3(a + 2)$
= ______
= ______

(b) $3(e + 3) + 4(e - 5)$
= ______
= ______

(c) $2(5p + q) + 4(3p - 7q)$
= ______
= ______

(d) $6(9y + 4z) + (y - 2z)$
= ______
= ______

2. **Expand and simplify:**

(a) $5y - 3(2y + 5)$
= ______
= ______

(b) $4k - 7(k - 8)$
= ______
= ______

(c) $9 - (h - 5)$
= ______
= ______

(d) $8g - (2g + 7)$
= ______
= ______

(e) $7(3v + 1) - 4(2v + 3)$
= ______
= ______

(f) $2(5s + 4t) - 3(3s - 2t)$
= ______
= ______

(g) $4(b + 6) - 3(b - 6)$
= ______
= ______

(h) $6y + 3z - (2y + 3z)$
= ______
= ______

(i) $8m(2m - 3n) - 5(2m^2 + 5n)$
= ______
= ______

(j) $3(2q - 5) - 4(q - 3)$
= ______
= ______

(k) $2(d + 4) - (d - 7)$
= ______
= ______

(l) $r(7r - 3) - (r^2 + 2r - 3)$
= ______
= ______

3. **Expand and simplify:**

(a) $3(2x - 1) + (4x + 3)$
= ______
= ______

(b) $2(u + 3) - 3(u + 2)$
= ______
= ______

(c) $5(2q - 5) + 3(2q - 1)$
= ______
= ______

(d) $7 - (2x + 3y - 9)$
= ______
= ______

(e) $2c(3c - 5) - (c^2 + 8c - 1)$
= ______
= ______

(f) $a(b + c) - b(a + b - c)$
= ______
= ______

EXPANDING AND SIMPLIFYING Worked Solutions

1. Expand and simplify:

(a) $4(a-6)+3(a+2)$

$= \mathbf{4a - 24 + 3a + 6}$ ☞ Expand both brackets, then collect like terms.

$= \mathbf{7a - 18}$

(b) $3(e+3)+4(e-5)$ ☞ $3 \times e + 3 \times 3 + 4 \times e - 4 \times 5$

$= \mathbf{3e + 9 + 4e - 20}$ — $3e$ $+9$ $+4e$ -20

$= \mathbf{7e - 11}$

(c) $2(5p+q)+4(3p-7q)$

$= \mathbf{10p + 2q + 12p - 28q}$

$= \mathbf{22p - 26q}$ ☞ $10p + 12p = 22p$, $2q - 28q = -26q$

(d) $6(9y+4z)+(y-2z)$

$= \mathbf{54y + 24z + y - 2z}$

$= \mathbf{55y + 22z}$

2. Expand and simplify:

(a) $5y-3(2y+5)$ ☞ The signs inside the bracket must change.

$= \mathbf{5y - 6y - 15}$

$= \mathbf{-y - 15}$

(b) $4k-7(k-8)$

$= \mathbf{4k - 7k + 56}$

$= \mathbf{-3k + 56}$ ☞ $56 - 3k$ is the same thing.

(c) $9-(h-5)$

$= \mathbf{9 - h + 5}$

$= \mathbf{14 - h}$ ☞ or $-h + 14$

(d) $8g-(2g+7)$

$= \mathbf{8g - 2g - 7}$

$= \mathbf{6g - 7}$

(e) $7(3v+1)-4(2v+3)$

$= \mathbf{21v + 7 - 8v - 12}$ ☞ Expand both brackets, watching the signs, then collect like terms.

$= \mathbf{13v - 5}$

(f) $2(5s+4t)-3(3s-2t)$

$= \mathbf{10s + 8t - 9s + 6t}$

$= \mathbf{s + 14t}$

(g) $4(b+6)-3(b-6)$

$= \mathbf{4b + 24 - 3b + 18}$ ☞ $4b + 24 - 3b + 18$

$= \mathbf{b + 42}$

(h) $6y + 3z - (2y + 3z)$
$= \mathbf{6y + 3z - 2y - 3z}$
$= \mathbf{4y}$ — $6y - 2y = 4y$, $3z - 3z = 0$

(i) $8m(2m - 3n) - 5(2m^2 + 5n)$
$= \mathbf{16m^2 - 24mn - 10m^2 - 25n}$ — Expand both sets of brackets,
$= \mathbf{6m^2 - 24mn - 25n}$ — collect like terms.

(j) $3(2q - 5) - 4(q - 3)$
$= \mathbf{6q - 15 - 4q + 12}$
$= \mathbf{2q - 3}$

(k) $2(d + 4) - (d - 7)$ — This is the same as $2(d + 4) - 1(d - 7)$.
$= \mathbf{2d + 8 - d + 7}$
$= \mathbf{d + 15}$

(l) $r(7r - 3) - (r^2 + 2r - 3)$ — All the signs in the second bracket will change.
$= \mathbf{7r^2 - 3r - r^2 - 2r + 3}$ — $7r^2 \; - 3r \; - r^2 \; - 2r \; + 3$
$= \mathbf{6r^2 - 5r + 3}$

3. Expand and simplify:

(a) $3(2x - 1) + (4x + 3)$ — Expand,
$= \mathbf{6x - 3 + 4x + 3}$ — collect like terms.
$= \mathbf{10x}$

(b) $2(u + 3) - 3(u + 2)$
$= \mathbf{2u + 6 - 3u - 6}$
$= \mathbf{-u}$ — $2u - 3u = -u$, $6 - 6 = 0$

(c) $5(2q - 5) + 3(2q - 1)$
$= \mathbf{10q - 25 + 6q - 3}$
$= \mathbf{16q - 28}$

(d) $7 - (2x + 3y - 9)$ — All the signs in the bracket will change.
$= \mathbf{7 - 2x - 3y + 9}$
$= \mathbf{16 - 2x - 3y}$

(e) $2c(3c - 5) - (c^2 + 8c - 1)$
$= \mathbf{6c^2 - 10c - c^2 - 8c + 1}$
$= \mathbf{5c^2 - 18c + 1}$

(f) $a(b + c) - b(a + b - c)$
$= \mathbf{ab + ac - ab - b^2 + bc}$ — ba is the same thing as ab.
$= \mathbf{ac - b^2 + bc}$

BINOMIAL PRODUCTS

Just as a bicycle has two wheels, a binomial has two terms.
The multiplication of two binomials is called a *binomial product*.

$(x + y)(x + 3)$ means that $x + 3$ must be multiplied by x and by y.

$$\begin{aligned}(x + y)(x + 3) &= x(x + 3) + y(x + 3)\\ &= x^2 + 3x + xy + 3y\end{aligned}$$

Example:

Find:

(a) $(a + 2)(a + 5)$

(b) $(3x + 4)(5x - 1)$

(c) $(8a - 3b)(2a + 5b)$

(d) $(3t - 1)(2t - 1)$

Solution:

(a) $$\begin{aligned}(a + 2)(a + 5) &= \mathbf{a(a + 5) + 2(a + 5)}\\ &= \mathbf{a^2 + 5a + 2a + 10}\\ &= \mathbf{a^2 + 7a + 10}\end{aligned}$$

(b) $$\begin{aligned}(3x + 4)(5x - 1) &= \mathbf{3x(5x - 1) + 4(5x - 1)}\\ &= \mathbf{15x^2 - 3x + 20x - 4}\\ &= \mathbf{15x^2 + 17x - 4}\end{aligned}$$

(c) $$\begin{aligned}(8a - 3b)(2a + 5b) &= \mathbf{8a(2a + 5b) - 3b(2a + 5b)}\\ &= \mathbf{16a^2 + 40ab - 6ab - 15b^2}\\ &= \mathbf{16a^2 + 34ab - 15b^2}\end{aligned}$$

(d) $$\begin{aligned}(3t - 1)(2t - 1) &= \mathbf{3t(2t - 1) - 1(2t - 1)}\\ &= \mathbf{6t^2 - 3t - 2t + 1}\\ &= \mathbf{6t^2 - 5t + 1}\end{aligned}$$

A binomial product can also be explained using areas:

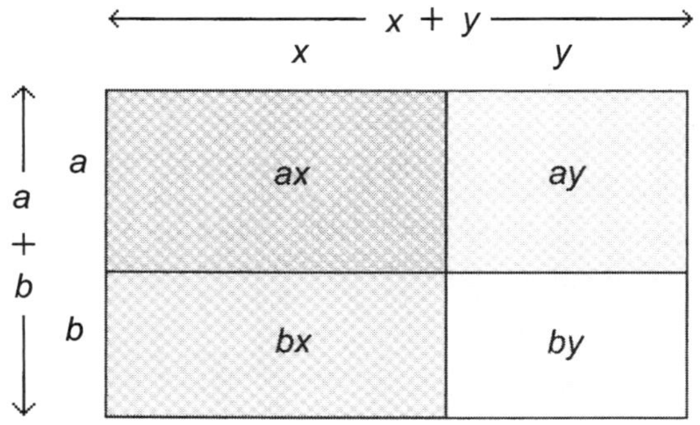

Each of the terms of the second binomial must be multiplied by each term of the first.

We can use this as a rule: $(a + b)(x + y) = ax + ay + bx + by$

Example:

Find:

(a) $(e + 4)(e + 7)$

(b) $(2z + 5)(3z - 1)$

(c) $(8p - 3q)(5p + 3q)$

(d) $(6 - y)(3y - 7)$

Solution:

(a) $$\begin{aligned}(e + 4)(e + 7) &= \mathbf{e^2 + 7e + 4e + 28}\\ &= \mathbf{e^2 + 11e + 28}\end{aligned}$$

(b) $$\begin{aligned}(2z + 5)(3z - 1) &= \mathbf{6z^2 - 2z + 15z - 5}\\ &= \mathbf{6z^2 + 13z - 5}\end{aligned}$$

(c) $$\begin{aligned}(8p - 3q)(5p + 3q) &= \mathbf{40p^2 + 24pq - 15pq - 9q^2}\\ &= \mathbf{40p^2 + 9pq - 9q^2}\end{aligned}$$

(d) $$\begin{aligned}(6 - y)(3y - 7) &= \mathbf{18y - 42 - 3y^2 + 7y}\\ &= \mathbf{-3y^2 + 25y - 42}\end{aligned}$$

1. **Expand and simplify:**

(a) $a(a + 2) + 4(a + 2)$
= ____________________
= ____________________

(b) $x(x - 4) + 3(x - 4)$
= ____________________
= ____________________

(c) $t(2t - 3u) + u(2t - 3u)$
= ____________________
= ____________________

(d) $m(m + 6) - 7(m + 6)$
= ____________________
= ____________________

(e) $3k(k + n) + 4(k + n)$
= ____________________
= ____________________

(f) $4q(2q - 9) - 3(2q - 9)$
= ____________________
= ____________________

2. **Complete:**

(a) $(a + 3)(a + 8) = a(a + 8) +$ __________
$= a^2 + 8a +$ __________
$= a^2 +$ __________

(b) $(x - 5)(x - 6) = x($______$) - 5($________$)$
= ____________________
= ____________________

3. **Find these binomial products, (using three steps as in Question 2):**

(a) $(a + 3)(a + 5)$
= ____________________
= ____________________
= ____________________

(b) $(p + q)(p + 4)$
= ____________________
= ____________________
= ____________________

(c) $(d + 6)(d - 2)$
= ____________________
= ____________________
= ____________________

(d) $(t - 8)(t + 3)$
= ____________________
= ____________________
= ____________________

(e) $(k - 6)(k - 4)$
= ____________________
= ____________________
= ____________________

(f) $(2x + 3)(x + 4)$
= ____________________
= ____________________
= ____________________

(g) $(3e - 2)(2f + 7)$
= ____________________
= ____________________
= ____________________

(h) $(6 - 5n)(4 - 3n)$
= ____________________
= ____________________
= ____________________

(i) $(5x + 2y)(3x - y)$
= ____________________
= ____________________
= ____________________

4. **Find these binomial products:**

(a) $(x + 5)(x + 9)$
= ____________________
= ____________________

(b) $(m + n)(p + q)$
= ____________________
= ____________________

(c) $(a + b)(a - 3)$
= ____________________
= ____________________

(d) $(4z + 1)(3z + 1)$
= ____________________
= ____________________

(e) $(8t - 3)(t - 3)$
= ____________________
= ____________________

(f) $(5g - 2)(2g + 9)$
= ____________________
= ____________________

(g) $(2c + 5)(c - 3)$
= ____________________
= ____________________

(h) $(7z - 1)(3z + 1)$
= ____________________
= ____________________

(i) $(q - 4)(3q - 4)$
= ____________________
= ____________________

20 BINOMIAL PRODUCTS Worked Solutions

1. Expand and simplify:

   (a) $a(a + 2) + 4(a + 2)$
   $= \mathbf{a^2 + 2a + 4a + 8}$
   $= \mathbf{a^2 + 6a + 8}$

   ☞ Remove both sets of brackets, then collect like terms.

   (b) $x(x - 4) + 3(x - 4)$
   $= \mathbf{x^2 - 4x + 3x - 12}$
   $= \mathbf{x^2 - x - 12}$

   (c) $t(2t - 3u) + u(2t - 3u)$
   $= \mathbf{2t^2 - 3tu + 2tu - 3u^2}$
   $= \mathbf{2t^2 - tu - 3u^2}$

   ☞ Remember, *ut* is the same thing as *tu*.

   (d) $m(m + 6) - 7(m + 6)$
   $= \mathbf{m^2 + 6m - 7m - 42}$
   $= \mathbf{m^2 - m - 42}$

   ☞ Be careful to change the sign in the second bracket.

   (e) $3k(k + n) + 4(k + n)$
   $= \mathbf{3k^2 + 3kn + 4k + 4n}$

   ☞ These are unlike terms, so this cannot be simplified.

   (f) $4q(2q - 9) - 3(2q - 9)$
   $= \mathbf{8q^2 - 36q - 6q + 27}$
   $= \mathbf{8q^2 - 42q + 27}$

   ☞ Be careful with the signs.

2. Complete:

   (a) $(a + 3)(a + 8)$
   $= a(a + 8) + \mathbf{3(a + 8)}$
   $= a^2 + 8a + \mathbf{3a + 24}$
   $= a^2 + \mathbf{11a + 24}$

   (b) $(x - 5)(x - 6)$
   $= \mathbf{x(x - 6) - 5(x - 6)}$
   $= \mathbf{x^2 - 6x - 5x + 30}$
   $= \mathbf{x^2 - 11x + 30}$

3. Find these binomial products, (using three steps as in Question 2):

   (a) $(a + 3)(a + 5)$
   $= \mathbf{a(a + 5) + 3(a + 5)}$
   $= \mathbf{a^2 + 5a + 3a + 15}$
   $= \mathbf{a^2 + 8a + 15}$

   ☞ The terms in the second bracket must be multiplied by each of the terms in the first bracket.

   (b) $(p + q)(p + 4)$
   $= \mathbf{p(p + 4) + q(p + 4)}$
   $= \mathbf{p^2 + 4p + pq + 4q}$

   ☞ These are all unlike terms, so this can't be simplified.

   (c) $(d + 6)(d - 2)$
   $= \mathbf{d(d - 2) + 6(d - 2)}$
   $= \mathbf{d^2 - 2d + 6d - 12}$
   $= \mathbf{d^2 + 4d - 12}$

   (d) $(t - 8)(t + 3)$
   $= \mathbf{t(t + 3) - 8(t + 3)}$
   $= \mathbf{t^2 + 3t - 8t - 24}$
   $= \mathbf{t^2 - 5t - 24}$

   ☞ The negative sign changes the sign in the bracket.

(e) $(k-6)(k-4)$
$= \mathbf{k(k-4)-6(k-4)}$
$= \mathbf{k^2-4k-6k+24}$
$= \mathbf{k^2-10k+24}$

(f) $(2x+3)(x+4)$
$= \mathbf{2x(x+4)+3(x+4)}$
$= \mathbf{2x^2+8x+3x+12}$
$= \mathbf{2x^2+11x+12}$

(g) $(3e-2)(2f+7)$
$= \mathbf{3e(2f+7)-2(2f+7)}$ ☞ Remember to change the sign.
$= \mathbf{6ef+21e-4f-14}$ ☞ These are unlike terms.

(h) $(6-5n)(4-3n)$
$= \mathbf{6(4-3n)-5n(4-3n)}$
$= \mathbf{24-18n-20n+15n^2}$ ☞ Be careful with the signs.
$= \mathbf{24-38n+15n^2}$

(i) $(5x+2y)(3x-y)$
$= \mathbf{5x(3x-y)+2y(3x-y)}$
$= \mathbf{15x^2-5xy+6xy-2y^2}$
$= \mathbf{15x^2+xy-2y^2}$ ☞ $1xy = xy$

4. Find these binomial products:

(a) $(x+5)(x+9)$
$= \mathbf{x^2+9x+5x+45}$ ☞ It is not necessary to write the first step (expansion), but it is not wrong to do so.
$= \mathbf{x^2+14x+45}$

(b) $(m+n)(p+q)$
$= \mathbf{mp+mq+np+nq}$ ☞ These are unlike terms and cannot be simplified.

(c) $(a+b)(a-3)$
$= \mathbf{a^2-3a+ab-3b}$ ☞ ba is the same thing as ab.

(d) $(4z+1)(3z+1)$
$= \mathbf{12z^2+4z+3z+1}$
$= \mathbf{12z^2+7z+1}$

(e) $(8t-3)(t-3)$
$= \mathbf{8t^2-24t-3t+9}$
$= \mathbf{8t^2-27t+9}$

(f) $(5g-2)(2g+9)$
$= \mathbf{10g^2+45g-4g-18}$
$= \mathbf{10g^2+41g-18}$

(g) $(2c+5)(c-3)$
$= \mathbf{2c^2-6c+5c-15}$ ☞ $-6c+5c=-1c$
$= \mathbf{2c^2-c-15}$

(h) $(7z-1)(3z+1)$
$= \mathbf{21z^2+7z-3z-1}$
$= \mathbf{21z^2+4z-1}$

(i) $(q-4)(3q-4)$
$= \mathbf{3q^2-4q-12q+16}$
$= \mathbf{3q^2-16q+16}$

22 SUM BY DIFFERENCE

When the sum of two terms is multiplied by their difference, we get a special result.
Consider the binomial product $(2x + 3y)(2x - 3y)$.
The two binomials have the same terms,
but in one they are added and in the other they are subtracted.

Notice that the two middle terms cancel each other out!

$$\begin{aligned}(2x + 3y)(2x - 3y) &= 4x^2 - 6xy + 6xy - 9y^2 \\ &= 4x^2 - 9y^2\end{aligned}$$

$$\begin{aligned}(a + b)(a - b) &= a^2 - ab + ab - b^2 \\ &= a^2 - b^2\end{aligned}$$

We use this as a rule: $\boxed{(a + b)(a - b) = a^2 - b^2}$

This rule only works when we multiply the **sum** of two terms **by** their **difference**.

Example:

Find:

(a) $(x + 5)(x - 5)$

(b) $(e - 8)(e + 8)$

(c) $(4 - g)(4 + g)$

(d) $(a + 1)(a - 1)$

Solution:

(a) $(x + 5)(x - 5) = \mathbf{x^2 - 5^2}$
$= \mathbf{x^2 - 25}$

(b) $(e - 8)(e + 8) = \mathbf{e^2 - 8^2}$
$= \mathbf{e^2 - 64}$

(c) $(4 - g)(4 + g) = \mathbf{4^2 - g^2}$
$= \mathbf{16 - g^2}$

(d) $(a + 1)(a - 1) = \mathbf{a^2 - 1^2}$
$= \mathbf{a^2 - 1}$

We must take care to square each part of every term.

Example:

Find:

(a) $(2k + 7m)(2k - 7m)$

(b) $(4p + 5q)(4p - 5q)$

(c) $(9p - 1)(9p + 1)$

(d) $(1 - 2k)(1 + 2k)$

(e) $(x^2 + yz)(x^2 - yz)$

(f) $(2m^3 + 3n^2)(2m^3 - 3n^2)$

Solution:

(a) $(2k + 7m)(2k - 7m) = \mathbf{(2k)^2 - (7m)^2}$
$= \mathbf{4k^2 - 49m^2}$

(b) $(4p + 5q)(4p - 5q) = \mathbf{(4p)^2 - (5q)^2}$
$= \mathbf{16p^2 - 25q^2}$

(c) $(9p - 1)(9p + 1) = \mathbf{(9p)^2 - 1^2}$
$= \mathbf{81p^2 - 1}$

(d) $(1 - 2k)(1 + 2k) = \mathbf{1^2 - (2k)^2}$
$= \mathbf{1 - 4k^2}$

(e) $(x^2 + yz)(x^2 - yz) = \mathbf{(x^2)^2 - (yz)^2}$
$= \mathbf{x^4 - y^2z^2}$

(f) $(2m^3 + 3n^2)(2m^3 - 3n^2) = \mathbf{(2m^3)^2 - (3n^2)^2}$
$= \mathbf{4m^6 - 9n^4}$

1. Complete:

(a) $(x + 4)(x - 4) = x^2 - \_\_\_\_\_\_^2$

(b) $(a + 7)(a - 7) = \_\_\_\_\_\_^2 - \_\_\_\_\_\_^2$

2. Find:

(a) $(x + 3)(x - 3)$ = ____________
= ____________

(b) $(a + 9)(a - 9)$ = ____________
= ____________

(c) $(t + 11)(t - 11)$ = ____________
= ____________

(d) $(y + 1)(y - 1)$ = ____________
= ____________

(e) $(m - 2)(m + 2)$ = ____________
= ____________

(f) $(p - q)(p + q)$ = ____________

(g) $(5n + 3)(5n - 3)$ = ____________
= ____________

(h) $(4u + 9)(4u - 9)$ = ____________
= ____________

(i) $(8t - u)(8t + u)$ = ____________
= ____________

(j) $(7z + 1)(7z - 1)$ = ____________
= ____________

(k) $(6e - 5)(6e + 5)$ = ____________
= ____________

(l) $(8g + 3h)(8g - 3h)$ = ____________
= ____________

(m) $(3 - 2p)(3 + 2p)$ = ____________
= ____________

(n) $(10y + 7z)(10y - 7z)$ = ____________
= ____________

(o) $(1 + d)(1 - d)$ = ____________
= ____________

(p) $(9r - 7s)(9r + 7s)$ = ____________
= ____________

3. Find:

(a) $(x + 6)(x - 6)$ = ____________

(b) $(t - 7)(t + 7)$ = ____________

(c) $(g + 1)(g - 1)$ = ____________

(d) $(y + z)(y - z)$ = ____________

(e) $(5 - m)(5 + m)$ = ____________

(f) $(2a + 3)(2a - 3)$ = ____________

(g) $(5x + 4y)(5x - 4y)$ = ____________

(h) $(1 - 9q)(1 + 9q)$ = ____________

4. Find:

(a) $(a^2 - 2a)(a^2 + 2a)$ = ____________
= ____________

(b) $(bc - a^2)(bc + a^2)$ = ____________
= ____________

(c) $(1 - x^3)(1 + x^3)$ = ____________
= ____________

(d) $(4n^2 + 5p^2)(4n^2 - 5p^2)$ = ____________
= ____________

24 SUM BY DIFFERENCE Worked Solutions

1. Complete:

(a) $(x + 4)(x - 4) = x^2 - \mathbf{4}^2$

(b) $(a + 7)(a - 7) = a^2 - \mathbf{7}^2$

2. Find:

(a) $(x + 3)(x - 3) = \mathbf{x^2 - 3^2}$
$= \mathbf{x^2 - 9}$

☞ The square of the first term minus the square of the second.

(b) $(a + 9)(a - 9) = \mathbf{a^2 - 9^2}$
$= \mathbf{a^2 - 81}$

(c) $(t + 11)(t - 11) = \mathbf{t^2 - 11^2}$
$= \mathbf{t^2 - 121}$

(d) $(y + 1)(y - 1) = \mathbf{y^2 - 1^2}$
$= \mathbf{y^2 - 1}$

(e) $(m - 2)(m + 2) = \mathbf{m^2 - 2^2}$
$= \mathbf{m^2 - 4}$

☞ $(m - 2)(m + 2)$ is the same thing as $(m + 2)(m - 2)$. The question is still a sum by difference.

(f) $(p - q)(p + q) = \mathbf{p^2 - q^2}$

☞ This can't be simplified.

(g) $(5n + 3)(5n - 3) = \mathbf{(5n)^2 - 3^2}$
$= \mathbf{25n^2 - 9}$

☞ Every part of each term must be squared. $(5n)^2 = 5n \times 5n = 25n^2$

(h) $(4u + 9)(4u - 9) = \mathbf{(4u)^2 - 9^2}$
$= \mathbf{16u^2 - 81}$

☞ $(4u)^2 = 4u \times 4u = 16u^2$

(i) $(8t - u)(8t + u) = \mathbf{(8t)^2 - u^2}$
$= \mathbf{64t^2 - u^2}$

(j) $(7z + 1)(7z - 1) = \mathbf{(7z)^2 - 1^2}$
$= \mathbf{49z^2 - 1}$

(k) $(6e - 5)(6e + 5) = \mathbf{(6e)^2 - 5^2}$
$= \mathbf{36e^2 - 25}$

(l) $(8g + 3h)(8g - 3h) = \mathbf{(8g)^2 - (3h)^2}$
$= \mathbf{64g^2 - 9h^2}$

☞ Every part of each term must be squared.

(m) $(3 - 2p)(3 + 2p) = \mathbf{3^2 - (2p)^2}$
$= \mathbf{9 - 4p^2}$

(n) $(10y + 7z)(10y - 7z) = \mathbf{(10y)^2 - (7z)^2}$
$= \mathbf{100y^2 - 49z^2}$

(o) $(1 + d)(1 - d) = \mathbf{1^2 - d^2}$
$= \mathbf{1 - d^2}$

(p) $(9r - 7s)(9r + 7s) = \mathbf{(9r)^2 - (7s)^2}$
$= \mathbf{81r^2 - 49s^2}$

3. Find:

(a) $(x + 6)(x - 6) = \mathbf{x^2 - 36}$ ☞ $x^2 - 6^2$

(b) $(t - 7)(t + 7) = \mathbf{t^2 - 49}$ ☞ $t^2 - 7^2$

(c) $(g + 1)(g - 1) = \mathbf{g^2 - 1}$

(d) $(y + z)(y - z) = \mathbf{y^2 - z^2}$

(e) $(5 - m)(5 + m) = \mathbf{25 - m^2}$ ☞ $5^2 - m^2$

(f) $(2a + 3)(2a - 3) = \mathbf{4a^2 - 9}$ ☞ Don't forget to square every part of each term. $(2a)^2 = 4a^2$

(g) $(5x + 4y)(5x - 4y) = \mathbf{25x^2 - 16y^2}$ ☞ $(5x)^2 - (4y)^2$

(h) $(1 - 9q)(1 + 9q) = \mathbf{1 - 81q^2}$ ☞ $1^2 - (9q)^2$

4. Find:

(a) $(a^2 - 2a)(a^2 + 2a) = \mathbf{(a^2)^2 - (2a)^2}$
$= \mathbf{a^4 - 4a^2}$

☞ $(a^2)^2 = a^2 \times a^2 = a^4$
(or when raising a power to a power, multiply the indices $(a^2)^2 = a^{2\times 2} = a^4$).

(b) $(bc - a^2)(bc + a^2) = \mathbf{(bc)^2 - (a^2)^2}$
$= \mathbf{b^2c^2 - a^4}$

☞ $(bc)^2 = bc \times bc = b^2c^2$

(c) $(1 - x^3)(1 + x^3) = \mathbf{1^2 - (x^3)^2}$
$= \mathbf{1 - x^6}$

☞ $(x^3)^2 = x^6$ The indices are multiplied.

(d) $(4n^2 + 5p^2)(4n^2 - 5p^2) = \mathbf{(4n^2)^2 - (5p^2)^2}$
$= \mathbf{16n^4 - 25p^4}$

☞ $(4n^2)^2 = 4n^2 \times 4n^2 = 16n^4$ $(5p^2)^2 = 5p^2 \times 5p^2 = 25p^4$

THE SQUARE OF A BINOMIAL

Finding the square of a binomial is a special case of a binomial product.

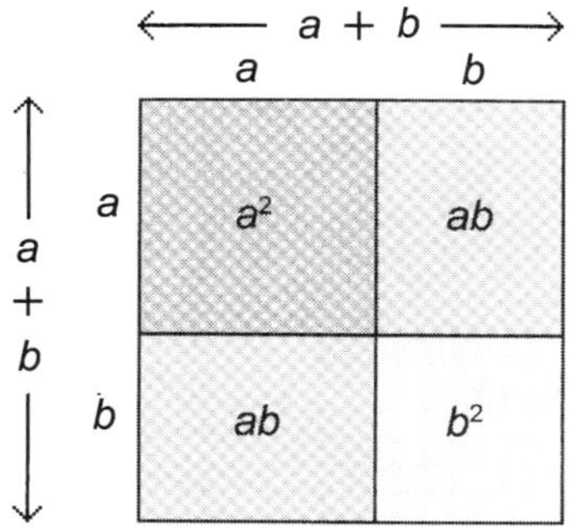

$$(a + b)^2 = (a + b)(a + b)$$
$$= a^2 + ab + ab + b^2$$
$$= a^2 + 2ab + b^2$$

We use the rules: $(a + b)^2 = a^2 + 2ab + b^2$ $(a - b)^2 = a^2 - 2ab + b^2$

Example:

Find:

(a) $(x + 7)^2$ **(b)** $(e - 6)^2$ **(c)** $(2p + 5)^2$ **(d)** $(4z - 1)^2$

Solution:

(a) $(x + 7)^2 = x^2 + 2 \times x \times 7 + 7^2$
$= x^2 + 14x + 49$

(b) $(e - 6)^2 = e^2 - 2 \times e \times 6 + 6^2$
$= e^2 - 12e + 36$

(c) $(2p + 5)^2 = (2p)^2 + 2 \times 2p \times 5 + 5^2$
$= 4p^2 + 20p + 25$

(d) $(4z - 1)^2 = (4z)^2 - 2 \times 4z \times 1 + 1^2$
$= 16z^2 - 8z + 1$

Note: We could do these questions just like any other binomial product:
$(3x + 8y)^2 = (3x + 8y)(3x + 8y)$
But learning these rules makes finding the square of a binomial quick and easy!

Example:

Find:

(a) $(3x + 8y)^2$ **(b)** $(9p - 7q)^2$ **(c)** $(x^2 + 3)^2$ **(d)** $(1 - 2xy)^2$

Solution:

(a) $(3x + 8y)^2 = (3x)^2 + 2 \times 3x \times 8y + (8y)^2$
$= 9x^2 + 48xy + 64y^2$

(b) $(9p - 7q)^2 = (9p)^2 - 2 \times 9p \times 7q + (7q)^2$
$= 81p^2 - 126pq + 49q^2$

(c) $(x^2 + 3)^2 = (x^2)^2 + 2 \times x^2 \times 3 + 3^2$
$= x^4 + 6x^2 + 9$

(d) $(1 - 2xy)^2 = 1^2 - 2 \times 1 \times 2xy + (2xy)^2$
$= 1 - 4xy + 4x^2y^2$

1. Complete:

(a) $(a + 5)^2 = a^2 + \_\_\_\_ \times a \times 5 + 5^2$

(b) $(e + 3)^2 = e^2 + 2 \times \_\_\_\_ \times \_\_\_\_ + 3^2$

(c) $(g + 8)^2 = g^2 + \_\_\_\_\_\_\_\_ + 8^2$

(d) $(p - 4)^2 = p^2 \_\_\_\_ 2 \times p \times 4 \_\_\_\_ 4^2$

(e) $(h - 1)^2 = h^2 - 2 \times h \times 1 + \_\_\_\_$

(f) $(3y + 2z)^2 = (\_\_\_\_)^2 + 2 \times \_\_\_\_ \times \_\_\_\_ + (\_\_\_\_)^2$

2. Find:

(a) $(x + 9)^2$ = ______________________
= _________________

(b) $(t + 11)^2$ = ______________________
= _________________

(c) $(n - 2)^2$ = ______________________
= _________________

(d) $(8 + k)^2$ = ______________________
= _________________

(e) $(h + 10)^2$ = ______________________
= _________________

(f) $(z - 1)^2$ = ______________________
= _________________

(g) $(2n + 7)^2$ = ______________________
= _________________

(h) $(3y - z)^2$ = ______________________
= _________________

(i) $(5t + 1)^2$ = ______________________
= _________________

(j) $(4b - 3)^2$ = ______________________
= _________________

(k) $(6 - 5v)^2$ = ______________________
= _________________

(l) $(3x + 2y)^2$ = ______________________
= _________________

(m) $(4e - 11f)^2$ = ______________________
= _________________

(n) $(p - 4q)^2$ = ______________________
= _________________

(o) $(3 - 7s)^2$ = ______________________
= _________________

(p) $(9m - 5n)^2$ = ______________________
= _________________

(q) $(12a - 5b)^2$ = ______________________
= _________________

(r) $(8c + 7d)^2$ = ______________________
= _________________

(s) $(x^2 + 2)^2$ = ______________________
= _________________

(t) $(e^2 - 6)^2$ = ______________________
= _________________

(u) $(ab + c)^2$ = ______________________
= _________________

(v) $(p^2 - qr)^2$ = ______________________
= _________________

(w) $(m^2 + n^2)^2$ = ______________________
= _________________

(x) $(1 - d^2)^2$ = ______________________
= _________________

28 THE SQUARE OF A BINOMIAL Worked Solutions

1. Complete:

(a) $(a+5)^2 = a^2 + \mathbf{2} \times a \times 5 + 5^2$ ☞ $(a+b)^2 = a^2 + 2ab + b^2$

(b) $(e+3)^2 = e^2 + 2 \times \boldsymbol{e} \times \mathbf{3} + 3^2$

(c) $(g+8)^2 = g^2 + \mathbf{2} \times \boldsymbol{g} \times \mathbf{8} + 8^2$

(d) $(p-4)^2 = p^2 - 2 \times p \times 4 + 4^2$ ☞ $(a-b)^2 = a^2 - 2ab + b^2$

(e) $(h-1)^2 = h^2 - 2 \times h \times 1 + \mathbf{1^2}$

(f) $(3y+2z)^2 = (\mathbf{3y})^2 + 2 \times \mathbf{3y} \times \mathbf{2z} + (\mathbf{2z})^2$

2. Find:

(a) $(x+9)^2 = \mathbf{x^2 + 2 \times x \times 9 + 9^2}$
$= \mathbf{x^2 + 18x + 81}$

(b) $(t+11)^2 = \mathbf{t^2 + 2 \times t \times 11 + 11^2}$
$= \mathbf{t^2 + 22t + 121}$

(c) $(n-2)^2 = \mathbf{n^2 - 2 \times n \times 2 + 2^2}$ ☞ Make sure the operation signs are correct.
$= \mathbf{n^2 - 4n + 4}$

(d) $(8+k)^2 = \mathbf{8^2 + 2 \times 8 \times k + k^2}$
$= \mathbf{64 + 16k + k^2}$

(e) $(h+10)^2 = \mathbf{h^2 + 2 \times h \times 10 + 10^2}$
$= \mathbf{h^2 + 20h + 100}$

(f) $(z-1)^2 = \mathbf{z^2 - 2 \times z \times 1 + 1^2}$ ☞ Be careful with the signs.
$= \mathbf{z^2 - 2z + 1}$

(g) $(2n+7)^2 = \mathbf{(2n)^2 + 2 \times 2n \times 7 + 7^2}$ ☞ Every part of the first term must be squared.
$= \mathbf{4n^2 + 28n + 49}$ $2n \times 2n = 4n^2$

(h) $(3y-z)^2 = \mathbf{(3y)^2 - 2 \times 3y \times z + z^2}$
$= \mathbf{9y^2 - 6yz + z^2}$ ☞ $(3y)^2 = 3y \times 3y = 9y^2$

(i) $(5t+1)^2 = \mathbf{(5t)^2 + 2 \times 5t \times 1 + 1^2}$
$= \mathbf{25t^2 + 10t + 1}$

(j) $(4b-3)^2 = \mathbf{(4b)^2 - 2 \times 4b \times 3 + 3^2}$
$= \mathbf{16b^2 - 24b + 9}$

(k) $(6 - 5v)^2 = \mathbf{6^2 - 2 \times 6 \times 5v + (5v)^2}$
$= \mathbf{36 - 60v + 25v^2}$

☞ Every part of the last term must be squared.

(l) $(3x + 2y)^2 = \mathbf{(3x)^2 + 2 \times 3x \times 2y + (2y)^2}$
$= \mathbf{9x^2 + 12xy + 4y^2}$

☞ $(3x)^2 = 3x \times 3x = 9x^2$ $(2y)^2 = 2y \times 2y = 4y^2$

(m) $(4e - 11f)^2 = \mathbf{(4e)^2 - 2 \times 4e \times 11f + (11f)^2}$
$= \mathbf{16e^2 - 88ef + 121f^2}$

☞ $(4e)^2 = 4e \times 4e = 16e^2$ $(11f)^2 = 11f \times 11f = 121f^2$

(n) $(p - 4q)^2 = \mathbf{p^2 - 2 \times p \times 4q + (4q)^2}$
$= \mathbf{p^2 - 8pq + 16q^2}$

(o) $(3 - 7s)^2 = \mathbf{3^2 - 2 \times 3 \times 7s + (7s)^2}$
$= \mathbf{9 - 42s + 49s^2}$

☞ Check that the signs are correct.

(p) $(9m - 5n)^2 = \mathbf{(9m)^2 - 2 \times 9m \times 5n + (5n)^2}$
$= \mathbf{81m^2 - 90mn + 25n^2}$

(q) $(12a - 5b)^2 = \mathbf{(12a)^2 - 2 \times 12a \times 5b + (5b)^2}$
$= \mathbf{144a^2 - 120ab + 25b^2}$

(r) $(8c + 7d)^2 = \mathbf{(8c)^2 + 2 \times 8c \times 7d + (7d)^2}$
$= \mathbf{64c^2 + 112cd + 49d^2}$

(s) $(x^2 + 2)^2 = \mathbf{(x^2)^2 + 2 \times x^2 \times 2 + 2^2}$
$= \mathbf{x^4 + 4x^2 + 4}$

☞ $(x^2)^2 = x^2 \times x^2 = x^4$

(t) $(e^2 - 6)^2 = \mathbf{(e^2)^2 - 2 \times e^2 \times 6 + 6^2}$
$= \mathbf{e^4 - 12e^2 + 36}$

(u) $(ab + c)^2 = \mathbf{(ab)^2 + 2 \times ab \times c + c^2}$
$= \mathbf{a^2b^2 + 2abc + c^2}$

☞ $(ab)^2 = ab \times ab = a^2b^2$

(v) $(p^2 - qr)^2 = \mathbf{(p^2)^2 - 2 \times p^2 \times qr + (qr)^2}$
$= \mathbf{p^4 - 2p^2qr + q^2r^2}$

(w) $(m^2 + n^2)^2 = \mathbf{(m^2)^2 + 2 \times m^2 \times n^2 + (n^2)^2}$
$= \mathbf{m^4 + 2m^2n^2 + n^4}$

(x) $(1 - d^2)^2 = \mathbf{1 - 2 \times 1 \times d^2 + (d^2)^2}$
$= \mathbf{1 - 2d^2 + d^4}$

MIXED PRODUCTS

We now know how to expand different types of products.

Example:

Find:

(a) $(x + 5)(x - 3)$

(b) $(h + 4)^2$

(c) $(8y + 5)(8y - 5)$

(d) $(6t - 1)^2$

(e) $(6x - 5)(2x - 3)$

(f) $(7 - b)(7 + b)$

Solution:

(a) $(x + 5)(x - 3) = x^2 - 3x + 5x - 15$
$= x^2 + 2x - 15$

(b) $(h + 4)^2 = h^2 + 2 \times h \times 4 + 4^2$
$= h^2 + 8h + 16$

(c) $(8y + 5)(8y - 5) = (8y)^2 - 5^2$
$= 64y^2 - 25$

(d) $(6t - 1)^2 = (6t)^2 - 2 \times 6t \times 1 + 1^2$
$= 36t^2 - 12t + 1$

(e) $(6x - 5)(2x - 3) = 12x^2 - 18x - 10x + 15$
$= 12x^2 - 28x + 15$

(f) $(7 - b)(7 + b) = 7^2 - b^2$
$= 49 - b^2$

We can add and subtract any combination of these products, remembering to follow the rules and be careful with the signs.

{Use brackets to help keep the signs correct!}

Example:

Expand and simplify:

(a) $(x + 5)(x + 3) + (x + 2)(x + 4)$

(b) $(h + 2)^2 + (h + 1)^2$

(c) $(3p + 7)(2p - 1) - (4p + 3)(2p + 1)$

(d) $(a + 9)^2 - (a + 3)(a - 3)$

Solution:

(a) $(x + 5)(x + 3) + (x + 2)(x + 4)$
$= x^2 + 3x + 5x + 15 + x^2 + 4x + 2x + 8$
$= 2x^2 + 14x + 23$

(b) $(h + 2)^2 + (h + 1)^2$
$= h^2 + 2 \times h \times 2 + 2^2 + h^2 + 2 \times h \times 1 + 1^2$
$= h^2 + 4h + 4 + h^2 + 2h + 1$
$= 2h^2 + 6h + 5$

(c) $(3p + 7)(2p - 1) - (4p + 3)(2p + 1)$
$= 6p^2 - 3p + 14p - 7 - (8p^2 + 4p + 6p + 3)$
$= 6p^2 + 11p - 7 - (8p^2 + 10p + 3)$
$= 6p^2 + 11p - 7 - 8p^2 - 10p - 3$
$= -2p^2 + p - 10$

(d) $(a + 9)^2 - (a + 3)(a - 3)$
$= a^2 + 2 \times a \times 9 + 9^2 - (a^2 - 3^2)$
$= a^2 + 18a + 81 - (a^2 - 9)$
$= a^2 + 18a + 81 - a^2 + 9$
$= 18a + 90$

1. **Expand these binomial products:**

(a) $(m + 7)(m + 2) =$ ____________
$=$ ____________
$=$ ____________

(b) $(2p - 9q)(p + 3q) =$ ____________
$=$ ____________
$=$ ____________

2. **Find these sum by differences:**

(a) $(g - 12)(g + 12) =$ ____________
$=$ ____________

(b) $(3x - 7y)(3x + 7y) =$ ____________
$=$ ____________

3. **Find these squares of binomials:**

(a) $(z - 6)^2 =$ ____________
$=$ ____________

(b) $(6t + 11u)^2 =$ ____________
$=$ ____________

4. **Find:**

(a) $(3x - 4)(3x + 4)$
$=$ ____________
$=$ ____________

(b) $(9y + 4)^2$
$=$ ____________
$=$ ____________

(c) $(4x + 1)(2x - 3)$
$=$ ____________
$=$ ____________

(d) $(8y - 7)(8y + 7)$
$=$ ____________
$=$ ____________

(e) $(4 - 3f)^2$
$=$ ____________
$=$ ____________

(f) $(5g + h)(5g - h)$
$=$ ____________
$=$ ____________

(g) $(3t - 10)(t - 10)$
$=$ ____________
$=$ ____________

(h) $(1 - e)^2$
$=$ ____________
$=$ ____________

5. **Expand and simplify:**

(a) $(x + 3)(x + 2) + (x + 4)(x + 5)$

____

____

(b) $6t - 2 - (t - 2)^2$

____

____

____

____

(c) $(k + 5)(k - 5) + (k + 5)^2$

____

____

____

(d) $(h - 8)(h - 9) - (h + 9)(h - 9)$

____

____

____

____

(e) $(3e - 5)^2 - (5e - 3)^2$

____

____

____

____

(f) $(7m - 4n)(3m - 2n) - (2m + n)(m - 3n)$

____

____

____

____

MIXED PRODUCTS Worked Solutions

1. Expand these binomial products:

   (a) $(m+7)(m+2) = \mathbf{m(m+2)+7(m+2)}$ ☞ It isn't necessary to write this line.

   $= \mathbf{m^2+2m+7m+14}$

   $= \mathbf{m^2+9m+14}$

   (b) $(2p-9q)(p+3q) = \mathbf{2p(p+3q)-9q(p+3q)}$

   $= \mathbf{2p^2+6pq-9pq-27q^2}$ ☞ qp is the same thing as pq.

   $= \mathbf{2p^2-3pq-27q^2}$

2. Find these sum by differences:

   (a) $(g-12)(g+12) = \mathbf{g^2-12^2}$

   $= \mathbf{g^2-144}$

   (b) $(3x-7y)(3x+7y) = \mathbf{(3x)^2-(7y)^2}$ ☞ Remember to square every part of each term.

   $= \mathbf{9x^2-49y^2}$

3. Find these squares of binomials:

   (a) $(z-6)^2 = \mathbf{z^2-2\times z\times 6+6^2}$ ☞ Check that the signs are correct.

   $= \mathbf{z^2-12z+36}$

   (b) $(6t+11u)^2 = \mathbf{(6t)^2+2\times 6t\times 11u+(11u)^2}$

   $= \mathbf{36t^2+132tu+121u^2}$

4. Find:

   (a) $(3x-4)(3x+4) = \mathbf{(3x)^2-4^2}$ ☞ This is a sum by difference.

   $= \mathbf{9x^2-16}$

   (b) $(9y+4)^2 = \mathbf{(9y)^2+2\times 9y\times 4+4^2}$ ☞ This is a square of a binomial.

   $= \mathbf{81y^2+72y+16}$

   (c) $(4x+1)(2x-3) = \mathbf{4x(2x-3)+1(2x-3)}$ ☞ This is a binomial product.

   $= \mathbf{8x^2-12x+2x-3}$ You could go straight to this step.

   $= \mathbf{8x^2-10x-3}$

   (d) $(8y-7)(8y+7) = \mathbf{(8y)^2-7^2}$ ☞ Sum by difference.

   $= \mathbf{64y^2-49}$

   (e) $(4-3f)^2 = \mathbf{4^2-2\times 4\times 3f+(3f)^2}$ ☞ Square of a binomial.

   $= \mathbf{16-24f+9f^2}$

   (f) $(5g+h)(5g-h) = \mathbf{(5g)^2-h^2}$ ☞ Sum by difference.

   $= \mathbf{25g^2-h^2}$

(g) $(3t - 10)(t - 10) = \mathbf{3t^2 - 30t - 10t + 100}$
$= \mathbf{3t^2 - 40t + 100}$

☞ Binomial product.

(h) $(1 - e)^2 = \mathbf{1^2 - 2 \times 1 \times e + e^2}$
$= \mathbf{1 - 2e + e^2}$

☞ Square of a binomial.

5. Expand and simplify:

(a) $(x + 3)(x + 2) + (x + 4)(x + 5)$
$= \mathbf{x^2 + 2x + 3x + 6 + x^2 + 5x + 4x + 20}$
$= \mathbf{2x^2 + 14x + 26}$

☞ This is the sum of two binomial products. First expand both binomials; then collect the like terms.

(b) $6t - 2 - (t - 2)^2 = \mathbf{6t - 2 - (t^2 - 2 \times t \times 2 + 2^2)}$
$= \mathbf{6t - 2 - (t^2 - 4t + 4)}$
$= \mathbf{6t - 2 - t^2 + 4t - 4}$
$= \mathbf{10t - 6 - t^2}$

☞ Use brackets.

The signs in the brackets change.

The answer could be written in a different order, provided the signs are correct. e.g. $-t^2 + 10t - 6$

(c) $(k + 5)(k - 5) + (k + 5)^2$
$= \mathbf{k^2 - 5^2 + k^2 + 2 \times k \times 5 + 5^2}$
$= \mathbf{k^2 - 25 + k^2 + 10k + 25}$
$= \mathbf{2k^2 + 10k}$

☞ Sum by difference plus the square of a binomial. Brackets are not necessary (because we are adding).

(d) $(h - 8)(h - 9) - (h + 9)(h - 9)$
$= \mathbf{h^2 - 9h - 8h + 72 - (h^2 - 9^2)}$
$= \mathbf{h^2 - 17h + 72 - (h^2 - 81)}$
$= \mathbf{h^2 - 17h + 72 - h^2 + 81}$
$= \mathbf{-17h + 153}$

(e) $(3e - 5)^2 - (5e - 3)^2$
$= \mathbf{(3e)^2 - 2 \times 3e \times 5 + 5^2 - ((5e)^2 - 2 \times 5e \times 3 + 3^2)}$
$= \mathbf{9e^2 - 30e + 25 - (25e^2 - 30e + 9)}$
$= \mathbf{9e^2 - 30e + 25 - 25e^2 + 30e - 9}$
$= \mathbf{-16e^2 + 16}$

(f) $(7m - 4n)(3m - 2n) - (2m + n)(m - 3n)$
$= \mathbf{21m^2 - 14mn - 12mn + 8n^2 - (2m^2 - 6mn + mn - 3n^2)}$
$= \mathbf{21m^2 - 26mn + 8n^2 - (2m^2 - 5mn - 3n^2)}$
$= \mathbf{21m^2 - 26mn + 8n^2 - 2m^2 + 5mn + 3n^2}$
$= \mathbf{19m^2 - 21mn + 11n^2}$

ALGEBRAIC FRACTIONS—ADDING AND SUBTRACTING

The rules for working with algebraic fractions are the same as those for other fractions. If a pair of fractions have a common denominator, to add (or subtract) we simply add (or subtract) the numerators.

Example:

Simplify:

(a) $\frac{2x}{7}+\frac{3x}{7}$ **(b)** $\frac{4a}{9b}+\frac{a}{9b}$ **(c)** $\frac{9}{y}-\frac{6}{y}$ **(d)** $\frac{t}{3}-\frac{2}{3}$

Solution:

(a) $\frac{2x}{7}+\frac{3x}{7}=\frac{5x}{7}$ (b) $\frac{4a}{9b}+\frac{a}{9b}=\frac{5a}{9b}$ (c) $\frac{9}{y}-\frac{6}{y}=\frac{3}{y}$ (d) $\frac{t}{3}-\frac{2}{3}=\frac{t-2}{3}$

> The answer should be simplified, if possible, but mixed numbers are not usually used with pronumerals.

Example:

Find:

(a) $\frac{5t}{9}+\frac{4t}{9}$ **(b)** $\frac{7e}{12}-\frac{e}{12}$ **(c)** $\frac{3t}{5}+\frac{3t}{5}$ **(d)** $\frac{9h}{4}-\frac{3h}{4}$

Solution:

(a) $\frac{5t}{9}+\frac{4t}{9}=\frac{9t}{9}$
$=t$

(b) $\frac{7e}{12}-\frac{e}{12}=\frac{6e}{12}$
$=\frac{e}{2}$ (or $\frac{1}{2}e$)

(c) $\frac{3t}{5}+\frac{3t}{5}=\frac{6t}{5}$

(d) $\frac{9h}{4}-\frac{3h}{4}=\frac{6h}{4}$
$=\frac{3h}{2}$

> To add or subtract fractions, we must have a common denominator.

Example:

Find:

(a) $\frac{4m}{7}+\frac{5m}{8}$ **(b)** $\frac{3a}{4}-\frac{a}{8}$ **(c)** $\frac{3}{x}-\frac{5}{y}$ **(d)** $\frac{e}{5}+\frac{1}{2}$

Solution:

(a) $\frac{4m}{7}+\frac{5m}{8}=\frac{32m}{56}+\frac{35m}{56}$
$=\frac{67m}{56}$

(b) $\frac{3a}{4}-\frac{a}{8}=\frac{6a}{8}-\frac{a}{8}$
$=\frac{5a}{8}$

(c) $\frac{3}{x}-\frac{5}{y}=\frac{3y}{xy}-\frac{5x}{xy}$
$=\frac{3y-5x}{xy}$

(d) $\frac{e}{5}+\frac{1}{2}=\frac{2e}{10}+\frac{5}{10}$
$=\frac{2e+5}{10}$

THINK:
7 and 8 both divide into what number?
56
What do we multiply 7 by to get 56?
8
(So we must also multiply 4*m* by 8.)
32*m*
What do we multiply 8 by to get 56?
7
(So we must also multiply 5*m* by 7.)
35*m*

1. Write as a single fraction:

(a) $\frac{a}{5}+\frac{a}{5}$ = ________ (d) $\frac{4y}{5}-\frac{3y}{5}$ = ________ (g) $\frac{2}{9n}+\frac{5}{9n}$ = ________

(b) $\frac{2d}{9}+\frac{5d}{9}$ = ________ (e) $\frac{8t}{15}-\frac{t}{15}$ = ________ (h) $\frac{a}{4}+\frac{3}{4}$ = ________

(c) $\frac{3k}{7}+\frac{k}{7}$ = ________ (f) $\frac{6}{x}+\frac{3}{x}$ = ________ (i) $\frac{7}{y}-\frac{x}{y}$ = ________

2. Add or subtract these fractions, simplifying the answer where possible:

(a) $\frac{h}{6}+\frac{h}{6}$ = ________ = ________ (c) $\frac{9k}{10}-\frac{7k}{10}$ = ________ = ________ (e) $\frac{11p}{12}-\frac{7p}{12}$ = ________ = ________

(b) $\frac{3w}{8}+\frac{5w}{8}$ = ________ = ________ (d) $\frac{5e}{9}+\frac{e}{9}$ = ________ = ________ (f) $\frac{5}{2m}+\frac{3}{2m}$ = ________ = ________

3. Add or subtract these fractions, leaving the answers as improper fractions:

(a) $\frac{3t}{5}+\frac{4t}{5}$ = ________ (c) $\frac{7p}{8}+\frac{5p}{8}$ = ________ = ________ (d) $\frac{11b}{4}-\frac{5b}{4}$ = ________ = ________

(b) $\frac{11g}{3}-\frac{g}{3}$ = ________

4. Complete:

(a) $\frac{3n}{5}=\frac{\square}{10}$ (b) $\frac{4x}{9}=\frac{\square}{27}$ (c) $\frac{6t}{11}=\frac{24t}{\square}$ (d) $\frac{7}{y}=\frac{\square}{6y}$ (e) $\frac{a}{b}=\frac{ac}{\square}$

5. Add or subtract:

(a) $\frac{3n}{5}+\frac{2n}{7}$ = ________ = ________ (c) $\frac{t}{3}+\frac{t}{4}$ = ________ = ________ (e) $\frac{a}{5}+\frac{b}{6}$ = ________ = ________

(b) $\frac{4e}{9}-\frac{5e}{12}$ = ________ = ________ (d) $\frac{7k}{8}-\frac{k}{2}$ = ________ = ________ (f) $\frac{3e}{4}-\frac{2e}{3}$ = ________ = ________

6. Add or subtract, simplifying the answer where possible:

(a) $\frac{c}{2}+\frac{5c}{6}$ = ________ = ________ = ________

(b) $\frac{9h}{10}-\frac{2h}{5}$ = ________ = ________ = ________

(c) $\frac{11g}{12}-\frac{g}{4}$ = ________ = ________ = ________

(d) $\frac{j}{2}+\frac{j}{8}$ = ________ = ________

(e) $2n-\frac{5n}{12}$ = ________ = ________

ALGEBRAIC FRACTIONS Worked Solutions (1)

1. Write as a single fraction:

(a) $\frac{a}{5} + \frac{a}{5} = \mathbf{\frac{2a}{5}}$ ☞ There is a common denominator. Add the numerators.

(b) $\frac{2d}{9} + \frac{5d}{9} = \mathbf{\frac{7d}{9}}$

(c) $\frac{3k}{7} + \frac{k}{7} = \mathbf{\frac{4k}{7}}$ (d) $\frac{4y}{5} - \frac{3y}{5} = \mathbf{\frac{y}{5}}$ (e) $\frac{8t}{15} - \frac{t}{15} = \mathbf{\frac{7t}{15}}$

(f) $\frac{6}{x} + \frac{3}{x} = \mathbf{\frac{9}{x}}$

(g) $\frac{2}{9n} + \frac{5}{9n} = \mathbf{\frac{7}{9n}}$

(h) $\frac{a}{4} + \frac{3}{4} = \mathbf{\frac{a+3}{4}}$ ☞ The terms in the numerator are unlike.

(i) $\frac{7}{y} - \frac{x}{y} = \mathbf{\frac{7-x}{y}}$

2. Add or subtract these fractions, simplifying the answer where possible:

(a) $\frac{h}{6} + \frac{h}{6} = \mathbf{\frac{2h}{6}}$

$= \mathbf{\frac{h}{3}}$ ☞ Dividing numerator and denominator by 2.

(b) $\frac{3w}{8} + \frac{5w}{8} = \mathbf{\frac{8w}{8}}$

$= \mathbf{w}$ ☞ Cancelling the 8s.

(c) $\frac{9k}{10} - \frac{7k}{10} = \mathbf{\frac{2k}{10}}$

$= \mathbf{\frac{k}{5}}$ ☞ Dividing both numerator and denominator by 2.

(d) $\frac{5e}{9} + \frac{e}{9} = \mathbf{\frac{6e}{9}}$

$= \mathbf{\frac{2e}{3}}$

(e) $\frac{11p}{12} - \frac{7p}{12} = \mathbf{\frac{4p}{12}}$

$= \mathbf{\frac{p}{3}}$

(f) $\frac{5}{2m} + \frac{3}{2m} = \mathbf{\frac{8}{2m}}$

$= \mathbf{\frac{4}{m}}$

3. Add or subtract these fractions, leaving the answers as improper fractions:

(a) $\frac{3t}{5} + \frac{4t}{5} = \mathbf{\frac{7t}{5}}$

(b) $\frac{11g}{3} - \frac{g}{3} = \mathbf{\frac{10g}{3}}$

(c) $\frac{7p}{8} + \frac{5p}{8} = \mathbf{\frac{12p}{8}}$ ☞ 4 divides into both numerator and denominator.

$= \mathbf{\frac{3p}{2}}$

(d) $\frac{11b}{4} - \frac{5b}{4} = \mathbf{\frac{6b}{4}}$

$= \mathbf{\frac{3b}{2}}$

4. Complete:

(a) $\frac{3n}{5} = \frac{\mathbf{6n}}{10}$ (c) $\frac{6t}{11} = \frac{24t}{\mathbf{44}}$ (e) $\frac{a}{b} = \frac{ac}{\mathbf{bc}}$

(b) $\frac{4x}{9} = \frac{\mathbf{12x}}{27}$ (d) $\frac{7}{y} = \frac{\mathbf{42}}{6y}$

5. Add or subtract:

(a) $\frac{3n}{5} + \frac{2n}{7} = \frac{21n}{35} + \frac{10n}{35}$

$= \frac{31n}{35}$

☞ The common denominator is 35. We must multiply the first fraction by 7 and the second fraction by 5.

(b) $\frac{4e}{9} - \frac{5e}{12} = \frac{16e}{36} - \frac{15e}{36}$

$= \frac{e}{36}$

☞ The lowest common denominator is 36. The first fraction is multiplied by 4 and the second fraction by 3.

(c) $\frac{t}{3} + \frac{t}{4} = \frac{4t}{12} + \frac{3t}{12}$

$= \frac{7t}{12}$

(d) $\frac{7k}{8} - \frac{k}{2} = \frac{7k}{8} - \frac{4k}{8}$

$= \frac{3k}{8}$

☞ The lowest common denominator is 8. The first fraction does not need to change.
Multiply the second fraction by 4.

(e) $\frac{a}{5} + \frac{b}{6} = \frac{6a}{30} + \frac{5b}{30}$

$= \frac{6a + 5b}{30}$

(f) $\frac{3e}{4} - \frac{2e}{3} = \frac{9e}{12} - \frac{8e}{12}$

$= \frac{e}{12}$

6. Add or subtract, simplifying the answer where possible:

(a) $\frac{c}{2} + \frac{5c}{6} = \frac{3c}{6} + \frac{5c}{6}$

$= \frac{8c}{6}$

$= \frac{4c}{3}$

☞ The lowest common denominator is 6.

☞ After dividing numerator and denominator by 2.

(b) $\frac{9h}{10} - \frac{2h}{5} = \frac{9h}{10} - \frac{4h}{10}$

$= \frac{5h}{10}$

$= \frac{h}{2}$

(c) $\frac{11g}{12} - \frac{g}{4} = \frac{11g}{12} - \frac{3g}{12}$

$= \frac{8g}{12}$

$= \frac{2g}{3}$

(d) $\frac{j}{2} + \frac{j}{8} = \frac{4j}{8} + \frac{j}{8}$

$= \frac{5j}{8}$

(e) $2n - \frac{5n}{12} = \frac{24n}{12} - \frac{5n}{12}$

$= \frac{19n}{12}$

38 ALGEBRAIC FRACTIONS — MULTIPLYING AND DIVIDING

To multiply fractions, multiply the numerators and multiply the denominators.

Example:

(a) $\frac{2}{m} \times \frac{7}{n}$ (b) $\frac{a}{5} \times \frac{b}{7}$ (c) $\frac{x}{3} \times \frac{x}{y}$ (d) $\frac{c}{d} \times \frac{e}{f}$ (e) $8 \times \frac{w}{9}$ (f) $\frac{2}{m} \times n$

Solution:

(a) $\frac{2}{m} \times \frac{7}{n} = \frac{14}{mn}$

(b) $\frac{a}{5} \times \frac{b}{7} = \frac{ab}{35}$

(c) $\frac{x}{3} \times \frac{x}{y} = \frac{x^2}{3y}$

(d) $\frac{c}{d} \times \frac{e}{f} = \frac{ce}{df}$

(e) $8 \times \frac{w}{9} = \frac{8}{1} \times \frac{w}{9}$
$= \frac{8w}{9}$

(f) $\frac{2}{m} \times n = \frac{2}{m} \times \frac{n}{1}$
$= \frac{2n}{m}$

If possible, cancel before multiplying.

Example:

(a) $\frac{4k}{5} \times \frac{3}{8}$ (b) $\frac{9t}{10} \times \frac{2t}{3}$ (c) $\frac{2p}{q} \times \frac{q}{5p}$ (d) $\frac{x^2}{4} \times \frac{2}{x^3}$

Solution:

(a) $\frac{4k}{5} \times \frac{3}{8} = \frac{{}^{1}\cancel{4}k}{5} \times \frac{3}{\cancel{8}_{2}}$
$= \frac{3k}{10}$

(b) $\frac{9t}{10} \times \frac{2t}{3} = \frac{{}^{3}\cancel{9}t}{\cancel{10}_{5}} \times \frac{{}^{1}\cancel{2}t}{\cancel{3}_{1}}$
$= \frac{3t^2}{5}$

(c) $\frac{2p}{q} \times \frac{q}{5p} = \frac{2\cancel{p}^{1}}{\cancel{q}_{1}} \times \frac{\cancel{q}^{1}}{5\cancel{p}_{1}}$
$= \frac{2}{5}$

(d) $\frac{x^2}{4} \times \frac{2}{x^3} = \frac{{}^{1}\cancel{x^2}}{{}_{2}\cancel{4}} \times \frac{\cancel{2}^{1}}{\cancel{x^3}_{x}}$
$= \frac{1}{2x}$

To divide by a fraction, multiply by its reciprocal.
That means turn the second fraction upside-down and then multiply!

Example:

(a) $\frac{a}{3} \div \frac{b}{2}$ (b) $\frac{5}{e} \div \frac{6}{g}$ (c) $4 \div \frac{b}{2a}$ (d) $\frac{x}{y} \div 2$ (e) $\frac{3}{x} \div \frac{7}{x}$ (f) $\frac{k^2}{3m} \div \frac{km^2}{9}$

Solution:

(a) $\frac{a}{3} \div \frac{b}{2} = \frac{a}{3} \times \frac{2}{b}$
$= \frac{2a}{3b}$

(b) $\frac{5}{e} \div \frac{6}{g} = \frac{5}{e} \times \frac{g}{6}$
$= \frac{5g}{6e}$

(c) $4 \div \frac{b}{2a} = \frac{4}{1} \times \frac{2a}{b}$
$= \frac{8a}{b}$

(d) $\frac{x}{y} \div 2 = \frac{x}{y} \times \frac{1}{2}$
$= \frac{x}{2y}$

(e) $\frac{3}{x} \div \frac{7}{x} = \frac{3}{x} \times \frac{x}{7}$
$= \frac{3}{{}_{1}\cancel{x}} \times \frac{\cancel{x}^{1}}{7}$
$= \frac{3}{7}$

(f) $\frac{k^2}{3m} \div \frac{km^2}{9} = \frac{k^2}{3m} \times \frac{9}{km^2}$
$= \frac{{}^{k}\cancel{k^2}}{{}_{1}\cancel{3}m} \times \frac{\cancel{9}^{3}}{{}_{1}\cancel{k}m^2}$
$= \frac{3k}{m^3}$

1. Find these products:

(a) $\frac{a}{5} \times \frac{b}{4} =$ ______ (e) $\frac{3}{7} \times \frac{a}{b} =$ ______ (i) $\frac{1}{2} \times \frac{m}{n} =$ ______ (m) $\frac{a}{3} \times \frac{a}{5} =$ ______

(b) $\frac{x}{7} \times \frac{y}{3} =$ ______ (f) $\frac{k}{3} \times \frac{4}{5} =$ ______ (j) $3 \times \frac{5}{c} =$ ______ (n) $\frac{5}{x} \times \frac{6}{x} =$ ______

(c) $\frac{2}{p} \times \frac{8}{q} =$ ______ (g) $\frac{1}{g} \times \frac{1}{h} =$ ______ (k) $\frac{d}{3} \times 4 =$ ______ (o) $\frac{a}{2} \times \frac{b}{3} \times \frac{c}{4}$

(d) $\frac{x}{5} \times \frac{4}{y} =$ ______ (h) $\frac{a}{b} \times \frac{c}{d} =$ ______ (l) $t \times \frac{5}{6} =$ ______ $=$ ______

2. Cancel before multiplying these fractions:

(a) $\frac{2}{a} \times \frac{b}{4} =$ ______ (e) $\frac{a}{7} \times \frac{7}{b} =$ ______ (i) $\frac{2a}{5} \times \frac{3a}{8} =$ ______ (m) $\frac{x^2}{3} \times \frac{2}{x} =$ ______

(b) $\frac{x}{6} \times \frac{9}{y} =$ ______ (f) $\frac{2e}{9} \times \frac{3f}{10} =$ ______ (j) $\frac{p}{q} \times \frac{q}{p} =$ ______ (n) $\frac{4t}{3} \times \frac{3}{8t} =$ ______

(c) $\frac{2p}{5} \times \frac{5p}{8} =$ ______ (g) $\frac{x}{8} \times \frac{2}{x} =$ ______ (k) $\frac{5ab}{8} \times \frac{2}{15a} =$ ______ (o) $\frac{8a^3}{b^2} \times \frac{b}{4a^2} =$ ______

(d) $\frac{3}{n} \times \frac{n}{4} =$ ______ (h) $\frac{3t}{2} \times \frac{1}{9} =$ ______ (l) $\frac{6p}{7} \times \frac{1}{3p} =$ ______

3. Divide:

(a) $\frac{a}{3} \div \frac{2}{x} =$ ______ (c) $\frac{7}{t} \div \frac{t}{3} =$ ______ (e) $\frac{2}{a} \div \frac{7}{b} =$ ______ (g) $\frac{4}{9} \div z =$ ______

$=$ ______ $=$ ______ $=$ ______ $=$ ______

(b) $\frac{1}{5} \div \frac{m}{n} =$ ______ (d) $\frac{r}{s} \div \frac{3}{5} =$ ______ (f) $\frac{5}{6} \div \frac{3}{n} =$ ______ (h) $12 \div \frac{c}{d} =$ ______

$=$ ______ $=$ ______ $=$ ______ $=$ ______

4. Divide, giving the answer in its simplest form:

(a) $\frac{x}{4} \div \frac{x}{5} =$ ______ (c) $\frac{2x}{y} \div \frac{4y}{x} =$ ______ (e) $\frac{c}{8} \div \frac{d}{12} =$ ______ (g) $\frac{a^4b}{3c} \div \frac{6a}{b^2c} =$ ______

$=$ ______ $=$ ______ $=$ ______ $=$ ______

$=$ ______ $=$ ______ $=$ ______ $=$ ______

(b) $\frac{2}{u} \div \frac{8}{z} =$ ______ (d) $\frac{3k}{8} \div 6 =$ ______ (f) $\frac{5}{x} \div \frac{2}{x} =$ ______

$=$ ______ $=$ ______ $=$ ______

$=$ ______ $=$ ______ $=$ ______

ALGEBRAIC FRACTIONS Worked Solutions (2)

1. Find these products:

(a) $\frac{a}{5} \times \frac{b}{4} = \frac{ab}{20}$ ☞ Multiply the numerators and multiply the denominators.

(b) $\frac{x}{7} \times \frac{y}{3} = \frac{xy}{21}$ (d) $\frac{x}{5} \times \frac{4}{y} = \frac{4x}{5y}$ (f) $\frac{k}{3} \times \frac{4}{5} = \frac{4k}{15}$ (h) $\frac{a}{b} \times \frac{c}{d} = \frac{ac}{bd}$

(c) $\frac{2}{p} \times \frac{8}{q} = \frac{16}{pq}$ (e) $\frac{3}{7} \times \frac{a}{b} = \frac{3a}{7b}$ (g) $\frac{1}{g} \times \frac{1}{h} = \frac{1}{gh}$ (i) $\frac{1}{2} \times \frac{m}{n} = \frac{m}{2n}$

(j) $3 \times \frac{5}{c} = \frac{3}{1} \times \frac{5}{c}$ ☞ 3 is the same thing as $\frac{3}{1}$.

$= \frac{15}{c}$

(k) $\frac{d}{3} \times 4 = \frac{d}{3} \times \frac{4}{1}$

$= \frac{4d}{3}$

(l) $t \times \frac{5}{6} = \frac{t}{1} \times \frac{5}{6}$ (m) $\frac{a}{3} \times \frac{a}{5} = \frac{a^2}{15}$ (n) $\frac{5}{x} \times \frac{6}{x} = \frac{30}{x^2}$ (o) $\frac{a}{2} \times \frac{b}{3} \times \frac{c}{4} = \frac{abc}{24}$

$= \frac{5t}{6}$

2. Cancel before multiplying these fractions:
(You don't need to repeat fractions. It is done here so that you can see the cancelling.)

(a) $\frac{2}{a} \times \frac{b}{4} = \frac{{}^{1}\cancel{2}}{a} \times \frac{b}{\cancel{4}_{2}}$ ☞ 2 divides into both 2 and 4.

$= \frac{b}{2a}$

(b) $\frac{x}{6} \times \frac{9}{y} = \frac{x}{{}_{2}\cancel{6}} \times \frac{\cancel{9}^{3}}{y}$ ☞ 3 divides into both 6 and 9.

$= \frac{3x}{2y}$

(c) $\frac{2p}{5} \times \frac{5p}{8} = \frac{{}^{1}\cancel{2}p}{{}_{1}\cancel{5}} \times \frac{{}^{1}\cancel{5}p}{\cancel{8}_{4}}$ ☞ We can cancel by 5 and by 2.

$= \frac{p^2}{4}$

(d) $\frac{3}{n} \times \frac{n}{4} = \frac{3}{{}_{1}\cancel{n}} \times \frac{\cancel{n}^{1}}{4}$ ☞ n divides into both numerator and denominator.

$= \frac{3}{4}$

(e) $\frac{a}{7} \times \frac{7}{b} = \frac{a}{{}_{1}\cancel{7}} \times \frac{\cancel{7}^{1}}{b}$ (g) $\frac{x}{8} \times \frac{2}{x} = \frac{{}^{1}\cancel{x}}{{}_{4}\cancel{8}} \times \frac{\cancel{2}^{1}}{\cancel{x}_{1}}$ (i) $\frac{2a}{5} \times \frac{3a}{8} = \frac{{}^{1}\cancel{2}a}{5} \times \frac{3a}{\cancel{8}_{4}}$ (k) $\frac{5ab}{8} \times \frac{2}{15a} = \frac{{}^{1}\cancel{5a}b}{{}_{4}\cancel{8}} \times \frac{\cancel{2}^{1}}{\cancel{15a}_{3}}$

$= \frac{a}{b}$ $= \frac{1}{4}$ $= \frac{3a^2}{20}$ $= \frac{b}{12}$

(f) $\frac{2e}{9} \times \frac{3f}{10} = \frac{{}^{1}\cancel{2}e}{{}_{3}\cancel{9}} \times \frac{{}^{1}\cancel{3}f}{\cancel{10}_{5}}$ (h) $\frac{3t}{2} \times \frac{1}{9} = \frac{{}^{1}\cancel{3}t}{2} \times \frac{1}{\cancel{9}_{3}}$ (j) $\frac{p}{q} \times \frac{q}{p} = \frac{{}^{1}\cancel{p}}{{}_{1}\cancel{q}} \times \frac{\cancel{q}^{1}}{\cancel{p}_{1}}$ (l) $\frac{6p}{7} \times \frac{1}{3p} = \frac{{}^{2}\cancel{6p}}{7} \times \frac{1}{\cancel{3p}_{1}}$

$= \frac{ef}{15}$ $= \frac{t}{6}$ $= 1$ $= \frac{2}{7}$

(m) $\frac{x^2}{3} \times \frac{2}{x} = \frac{{}^{x}\cancel{x^2}}{3} \times \frac{2}{\cancel{x}_1}$

$= \frac{2x}{3}$

(n) $\frac{4t}{3} \times \frac{3}{8t} = \frac{{}^{1}\cancel{4t}}{{}_{1}\cancel{3}} \times \frac{\cancel{3}^{1}}{\cancel{8t}_{2}}$

$= \frac{1}{2}$

(o) $\frac{8a^3}{b^2} \times \frac{b}{4a^2} = \frac{{}^{2a}\cancel{8a^3}}{{}_{b}\cancel{b^2}} \times \frac{\cancel{b}^{1}}{\cancel{4a^2}_{1}}$

$= \frac{2a}{b}$

☞ Both a^2 and b divide into the numerator and the denominator.

3. Divide:

(a) $\frac{a}{3} \div \frac{2}{x} = \frac{a}{3} \times \frac{x}{2}$

$= \frac{ax}{6}$

☞ To divide by a fraction, multiply by its reciprocal.

(b) $\frac{1}{5} \div \frac{m}{n} = \frac{1}{5} \times \frac{n}{m}$

$= \frac{n}{5m}$

(c) $\frac{7}{t} \div \frac{t}{3} = \frac{7}{t} \times \frac{3}{t}$

$= \frac{21}{t^2}$

(d) $\frac{r}{s} \div \frac{3}{5} = \frac{r}{s} \times \frac{5}{3}$

$= \frac{5r}{3s}$

(e) $\frac{2}{a} \div \frac{7}{b} = \frac{2}{a} \times \frac{b}{7}$

$= \frac{2b}{7a}$

(f) $\frac{5}{6} \div \frac{3}{n} = \frac{5}{6} \times \frac{n}{3}$

$= \frac{5n}{18}$

(g) $\frac{4}{9} \div z = \frac{4}{9} \times \frac{1}{z}$

$= \frac{4}{9z}$

(h) $12 \div \frac{c}{d} = 12 \times \frac{d}{c}$

$= \frac{12d}{c}$

4. Divide, giving the answer in its simplest form:

(a) $\frac{x}{4} \div \frac{x}{5} = \frac{x}{4} \times \frac{5}{x}$

☞ To divide by a fraction, multiply by its reciprocal.

$= \frac{{}^{1}\cancel{x}}{4} \times \frac{5}{\cancel{x}_{1}}$

$= \frac{5}{4}$

☞ This improper fraction does not contain pronumerals, so it is simplified.

$= 1\frac{1}{4}$

(b) $\frac{2}{u} \div \frac{8}{z} = \frac{2}{u} \times \frac{z}{8}$

$= \frac{{}^{1}\cancel{2}}{u} \times \frac{z}{\cancel{8}_{4}}$

$= \frac{z}{4u}$

(c) $\frac{2x}{y} \div \frac{4y}{x} = \frac{2x}{y} \times \frac{x}{4y}$

$= \frac{{}^{1}\cancel{2}x}{y} \times \frac{x}{{}_{2}\cancel{4}y}$

$= \frac{x^2}{2y^2}$

(d) $\frac{3k}{8} \div 6 = \frac{3k}{8} \times \frac{1}{6}$

$= \frac{{}^{1}\cancel{3}k}{8} \times \frac{1}{\cancel{6}_{2}}$

$= \frac{k}{16}$

(e) $\frac{c}{8} \div \frac{d}{12} = \frac{c}{8} \times \frac{12}{d}$

$= \frac{c}{{}_{2}\cancel{8}} \times \frac{\cancel{12}^{3}}{d}$

$= \frac{3c}{2d}$

(f) $\frac{5}{x} \div \frac{2}{x} = \frac{5}{x} \times \frac{x}{2}$

$= \frac{5}{{}_{1}\cancel{x}} \times \frac{\cancel{x}^{1}}{2}$

$= \frac{5}{2}$

$= 2\frac{1}{2}$

(g) $\frac{a^4b}{3c} \div \frac{6a}{b^2c} = \frac{a^4b}{3c} \times \frac{b^2c}{6a}$

$= \frac{{}^{a^3}\cancel{a^4}b}{3\cancel{c}_{1}} \times \frac{b^2\cancel{c}^{1}}{6\cancel{a}_{1}}$

$= \frac{a^3b^3}{18}$

SUBSTITUTION REVIEW

REMEMBER TO:

- write the expression;
- replace the pronumeral with a number;
- find the answer.

Example:

(a) If $a = -2$, find the value of $3a^2$

(b) When $m = -4$, $2m^2 - 3m = ?$

Solution:

(a) If $a = -2$, $3a^2 = 3 \times (-2)^2$
$= 3 \times 4$
$= 12$

(b) When $m = -4$, $2m^2 - 3m = 2 \times (-4)^2 - 3 \times -4$
$= 2 \times 16 + 12$
$= 32 + 12$
$= 44$

> You might not need to write out all these steps.
> If using a calculator, make sure you know how to use it properly so that you always get the right answer!

Example:

(a) If $p = 5$ and $q = 7$, find the value of $(p - q)^2$

(b) When $d = -1$, $4d^2 - d =$?

Solution:

(a) When $p = 5$ and $q = 7$,
$(p - q)^2 = (5 - 7)^2$
$= (-2)^2$
$= 4$

(b) When $d = -1$,
$4d^2 - d = 4 \times (-1)^2 - (-1)$
$= 4 \times 1 + 1$
$= 5$

Example:

If $a = 3$ and $b = -5$, the value of $(a + b)^3$ is ?

Solution:

When $a = 3$ and $b = -5$, $(a + b)^3 = (3 + (-5))^3$
$= (-2)^3$
$= -8$

> Completing a table of values simply involves substituting different values into the same expression.

Example:

Given $y = x - x^3$, draw up a table of values using the values $-3, -2, -1, 0, 1, 2, 3$ for x.

Solution:

x	-3	-2	-1	0	1	2	3
y	**24**	**6**	**0**	**0**	**0**	**-6**	**-24**

If $x = -3$, $x - x^3 = -3 - (-3)^3$
$= -3 - (-27)$
$= -3 + 27$
$= 24$

1. Find the value of $8 - m$, if:
   (a) $m = 2$
   (b) $m = -4$
   (c) $m = 13$

2. Find the value of $2k^2$, if:
   (a) $k = 3$
   (b) $k = -1$
   (c) $k = -4$

3. Find the value of:
   (a) $3t + 9$, if $t = 4$
   (b) $5a - 9$, if $a = -2$
   (c) $-2x^2$ if $x = -3$
   (d) $5x^2 + 1$, if $x = 2$
   (e) $h^2 - h$, if $h = -1$
   (f) $6 - k^2$, if $k = 3$
   (g) $(x + y)^2$, if $x = 3$ and $y = -2$
   (h) $c^2 + d^2$, if $c = 8$ and $d = -8$
   (i) $(a - b)^2$, if $a = 6$ and $b = 9$
   (j) $p^2 - pq$, if $p = -7$ and $q = 5$

4. Complete these tables of values:
   (a) $y = x^2 + x$

x	–3	–2	–1	0	1	2	3
y							

(b) $y = 3 - x^2$

x	–3	–2	–1	0	1	2	3
y							

(c) $y = x^3$

x	–3	–2	–1	0	1	2	3
y							

(d) $y = 2x^2 - 5x$

x	–3	–2	–1	0	1	2	3
y							

SUBSTITUTION REVIEW Worked Solutions

1. Find the value of $8 - m$, if:
   (a) $m = 2$

   $8 - m = \mathbf{8 - 2}$ ☞ Replace m with 2.
   $= \mathbf{6}$

   (b) $m = -4$

   $8 - m = \mathbf{8 - (-4)}$
   $= \mathbf{8 + 4}$
   $= \mathbf{12}$

   (c) $m = 13$

   $8 - m = \mathbf{8 - 13}$
   $= \mathbf{-5}$

2. Find the value of $2k^2$, if:
   (a) $k = 3$

   $2k^2 = \mathbf{2 \times 3^2}$ ☞ $2k^2$ means $2 \times k^2$
   $= \mathbf{2 \times 9}$
   $= \mathbf{18}$

   (b) $k = -1$

   $2k^2 = \mathbf{2 \times (-1)^2}$
   $= \mathbf{2 \times 1}$
   $= \mathbf{2}$

   (c) $k = -4$

   $2k^2 = \mathbf{2 \times (-4)^2}$
   $= \mathbf{2 \times 16}$
   $= \mathbf{32}$

3. Find the value of:
   (a) $3t + 9$, if $t = 4$

   $3t + 9 = \mathbf{3 \times 4 + 9}$
   $= \mathbf{12 + 9}$
   $= \mathbf{21}$

   (b) $5a - 9$, if $a = -2$

   $5a - 9 = \mathbf{5 \times -2 - 9}$
   $= \mathbf{-10 - 9}$
   $= \mathbf{-19}$

   (c) $-2x^2$ if $x = -3$

   $-2x^2 = \mathbf{-2 \times (-3)^2}$ ☞ $-2x^2 = -2 \times x^2$
   $= \mathbf{-2 \times 9}$
   $= \mathbf{-18}$

   (d) $5x^2 + 1$, if $x = 2$

   $5x^2 + 1 = \mathbf{5 \times 2^2 + 1}$
   $= \mathbf{5 \times 4 + 1}$
   $= \mathbf{20 + 1}$
   $= \mathbf{21}$

   (e) $h^2 - h$, if $h = -1$

   $h^2 - h = \mathbf{(-1)^2 - (-1)}$ ☞ Every h must be replaced by –1.
   $= \mathbf{1 + 1}$
   $= \mathbf{2}$

(f) $6 - k^2$, if $k = 3$

$6 - k^2 = \mathbf{6 - 3^2}$
$= \mathbf{6 - 9}$
$= \mathbf{-3}$

(g) $(x + y)^2$, if $x = 3$ and $y = -2$

$(x + y)^2 = \mathbf{(3 + (-2))^2}$
$= \mathbf{(3 - 2)^2}$
$= \mathbf{1^2}$
$= \mathbf{1}$

☞ Replace x with 3 and y with –2.

(h) $c^2 + d^2$, if $c = 8$ and $d = -8$

$c^2 + d^2 = \mathbf{8^2 + (-8)^2}$
$= \mathbf{64 + 64}$
$= \mathbf{128}$

(i) $(a - b)^2$, if $a = 6$ and $b = 9$

$(a - b)^2 = \mathbf{(6 - 9)^2}$
$= \mathbf{(-3)^2}$
$= \mathbf{9}$

(j) $p^2 - pq$, if $p = -7$ and $q = 5$

$p^2 - pq = \mathbf{(-7)^2 - (-7 \times 5)}$
$= \mathbf{49 - (-35)}$
$= \mathbf{49 + 35}$
$= \mathbf{84}$

☞ $p^2 - pq = (-7)^2 - (-7) \times 5$ is the same thing.

4. Complete these tables of values:

(a) $y = x^2 + x$

x	–3	–2	–1	0	1	2	3
y	**6**	**2**	**0**	**0**	**2**	**6**	**12**

☞ When $x = -3$, $y = \mathbf{(-3)^2 + (-3)}$
$= \mathbf{9 - 3}$
$= \mathbf{6}$

(b) $y = 3 - x^2$

x	–3	–2	–1	0	1	2	3
y	**–6**	**–1**	**2**	**3**	**2**	**–1**	**–6**

☞ When $x = -3$, $y = \mathbf{3 - (-3)^2}$
$= \mathbf{3 - 9}$
$= \mathbf{-6}$

(c) $y = x^3$

x	–3	–2	–1	0	1	2	3
y	**–27**	**–8**	**–1**	**0**	**1**	**8**	**27**

☞ When $x = -3$, $y = \mathbf{(-3)^3}$
$= \mathbf{-27}$

(d) $y = 2x^2 - 5x$

x	–3	–2	–1	0	1	2	3
y	**33**	**18**	**7**	**0**	**–3**	**–2**	**3**

☞ When $x = -3$, $2x^2 - 5x = \mathbf{2 \times (-3)^2 - 5 \times -3}$
$= \mathbf{2 \times 9 - (-15)}$
$= \mathbf{18 + 15}$
$= \mathbf{33}$

FUNCTION NOTATION

A function of x is simply a rule that uses the pronumeral x.
Although x can have different values, for each value of x there is only one possible value of the function. The short way to write a function of x is $f(x)$. We say "f of x".

Example:

Given $f(x) = 4x + 3$, find $f(2)$

Functions work like me!
Put in a value and get an answer.

Solution:

$f(x) = 4x + 3$ This is the function, or rule.

$f(2) = 4 \times 2 + 3$ $f(2)$ means replace x in the function by 2.

$= 8 + 3$

$= 11$

Example:

Given $f(x) = 5x - 1$, find

(a) $f(2)$ (b) $f(6)$ (c) $f(-1)$ (d) $f(0)$ (e) $f(11)$ (f) $f(-6)$

Solution:

(a) $f(x) = 5x - 1$
$f(2) = 5 \times 2 - 1$
$= 10 - 1$
$= 9$

(b) $f(x) = 5x - 1$
$f(6) = 5 \times 6 - 1$
$= 30 - 1$
$= 29$

(c) $f(x) = 5x - 1$
$f(-1) = 5 \times -1 - 1$
$= -5 - 1$
$= -6$

(d) $f(x) = 5x - 1$
$f(0) = 5 \times 0 - 1$
$= 0 - 1$
$= -1$

(e) $f(x) = 5x - 1$
$f(11) = 5 \times 11 - 1$
$= 55 - 1$
$= 54$

(f) $f(x) = 5x - 1$
$f(-6) = 5 \times -6 - 1$
$= -30 - 1$
$= -31$

We must be careful to substitute the value of the pronumeral every time it appears in the function.

Example:

Given: (a) $f(x) = x^2 + x$, find $f(4)$ (b) $f(x) = 3x^2 - 7x + 2$, find $f(-3)$

Solution:

(a) $f(x) = x^2 + x$
$f(4) = 4^2 + 4$
$= 16 + 4$
$= 20$

(b) $f(x) = 3x^2 - 7x + 2$
$f(-3) = 3 \times (-3)^2 - 7 \times -3 + 2$
$= 27 + 21 + 2$
$= 50$

The most usual notation for a function is $f(x)$, a function of x.
But $f(y)$ is still a function, a function of y.
$h(t)$ is also a function, a function of t.
We can use any pronumerals to show a function.

Example:

Given: (a) $f(t) = t - t^2$, find $f(3)$ (b) $g(z) = 9 - 2z$, find $g(-2)$

Solution:

(a) $f(t) = t - t^2$
$f(3) = 3 - 3^2$
$= 3 - 9$
$= -6$

(b) $g(z) = 9 - 2z$
$g(-2) = 9 - 2 \times -2$
$= 9 + 4$
$= 13$

1. **Complete:**

(a) $f(x) = 3x + 2$
$f(4) =$ ________
$=$ ________

(b) $f(x) = x - 9$
$f(2) =$ ________
$=$ ________

(c) $f(x) = 5x + 4$
$f(6) =$ ________
$=$ ________

(d) $f(x) = 8 - x$
$f(5) =$ ________
$=$ ________

(e) $f(x) = 2x - 7$
$f(1) =$ ________
$=$ ________

(f) $f(x) = 12 - 4x$
$f(3) =$ ________
$=$ ________

(g) $f(x) = x + 7$
$f(12) =$ ________
$=$ ________

(h) $f(x) = 4x + 5$
$f(-2) =$ ________
$=$ ________

(i) $f(x) = -3x + 8$
$f(0) =$ ________
$=$ ________

(j) $f(x) = 5x - 9$
$f(-7) =$ ________
$=$ ________

(k) $f(x) = -6x$
$f(-4) =$ ________
$=$ ________

(l) $f(x) = x$
$f(3) =$ ________

2. **Given:**

(a) $f(x) = x + 2$, find $f(8)$
____
____
____

(b) $f(x) = 3x - 1$, find $f(-3)$
____
____
____

(c) $f(x) = 4 - x$, find $f(-5)$
____
____
____

3. **Given $f(x) = 4x - 7$, find:**

(a) $f(5)$
____
____
____

(b) $f(1)$
____
____
____

(c) $f(-2)$
____
____
____

4. **Complete:**

(a) $f(x) = x^2 + x$
$f(7) =$ ________
$=$ ________
$=$ ________

(b) $f(x) = x^2 + 5x$
$f(-2) =$ ________
$=$ ________
$=$ ________

(c) $f(x) = 2x^2 - 6x$
$f(5) =$ ________
$=$ ________
$=$ ________

(d) $f(x) = x^2 - x - 1$
$f(3) =$ ________
$=$ ________
$=$ ________

(e) $f(x) = 4x - x^2$
$f(-1) =$ ________
$=$ ________
$=$ ________

(f) $f(x) = x^3 + x^2 + x$
$f(-2) =$ ________
$=$ ________
$=$ ________

(g) $f(t) = t^2$
$f(5) =$ ________
$=$ ________
$=$ ________

(h) $g(y) = 5y - 1$
$g(4) =$ ________
$=$ ________
$=$ ________

(i) $h(b) = 3b^2 - b$
$h(3) =$ ________
$=$ ________
$=$ ________

FUNCTION NOTATION Worked Solutions

1. Complete:

(a) $f(x) = 3x + 2$

$f(4) = \mathbf{3 \times 4 + 2}$
$= \mathbf{12 + 2}$
$= \mathbf{14}$

☞ This is the function or rule.
$f(4)$ means replace every x in the rule with 4.

(b) $f(x) = x - 9$

$f(2) = \mathbf{2 - 9}$
$= \mathbf{-7}$

☞ This is the function of x.
$f(2)$ means replace every x in the function with 2.

(c) $f(x) = 5x + 4$

$f(6) = \mathbf{5 \times 6 + 4}$
$= \mathbf{30 + 4}$
$= \mathbf{34}$

☞ This is the function of x.
Replace x with 6.

(d) $f(x) = 8 - x$

$f(5) = \mathbf{8 - 5}$
$= \mathbf{3}$

☞ Replace every x with 5.

(e) $f(x) = 2x - 7$

$f(1) = \mathbf{2 \times 1 - 7}$
$= \mathbf{2 - 7}$
$= \mathbf{-5}$

(f) $f(x) = 12 - 4x$

$f(3) = \mathbf{12 - 4 \times 3}$
$= \mathbf{12 - 12}$
$= \mathbf{0}$

(g) $f(x) = x + 7$

$f(12) = \mathbf{12 + 7}$
$= \mathbf{19}$

(h) $f(x) = 4x + 5$

$f(-2) = \mathbf{4 \times -2 + 5}$
$= \mathbf{-8 + 5}$
$= \mathbf{-3}$

(i) $f(x) = -3x + 8$

$f(0) = \mathbf{-3 \times 0 + 8}$
$= \mathbf{0 + 8}$
$= \mathbf{8}$

(j) $f(x) = 5x - 9$

$f(-7) = \mathbf{5 \times -7 - 9}$
$= \mathbf{-35 - 9}$
$= \mathbf{-44}$

(k) $f(x) = -6x$

$f(-4) = \mathbf{-6 \times -4}$
$= \mathbf{24}$

(l) $f(x) = x$

$f(3) = \mathbf{3}$

☞ This is the function.
Replace every x with 3.

2. Given:

(a) $f(x) = x + 2$, find $f(8)$

$\boldsymbol{f(x) = x + 2}$
$\boldsymbol{f(8) = 8 + 2}$
$= \mathbf{10}$

☞ First write the function,
then replace every x with 8.

(b) $f(x) = 3x - 1$, find $f(-3)$

$\boldsymbol{f(x) = 3x - 1}$
$\boldsymbol{f(-3) = 3 \times -3 - 1}$
$= \mathbf{-9 - 1}$
$= \mathbf{-10}$

(c) $f(x) = 4 - x$, find $f(-5)$

$\boldsymbol{f(x) = 4 - x}$
$\boldsymbol{f(-5) = 4 - (-5)}$
$= \mathbf{4 + 5}$
$= \mathbf{9}$

3. Given $f(x) = 4x - 7$, find:

(a) $f(5)$

$\mathbf{f(x) = 4x - 7}$ ☞ This is the function of *x*.

$\mathbf{f(5) = 4 \times 5 - 7}$ In this case, replace *x* with 5.

$\mathbf{= 20 - 7}$

$\mathbf{= 13}$

(b) $f(1)$

$\mathbf{f(x) = 4x - 7}$ ☞ We use the same function.

$\mathbf{f(1) = 4 \times 1 - 7}$ In this case, replace *x* with 1.

$\mathbf{= 4 - 7}$

$\mathbf{= -3}$

(c) $f(-2)$

$\mathbf{f(x) = 4x - 7}$

$\mathbf{f(-2) = 4 \times -2 - 7}$ ☞ In this case, replace *x* with –2.

$\mathbf{= -8 - 7}$

$\mathbf{= -15}$

4. Complete:

(a) $f(x) = x^2 + x$ ☞ Our function uses *x* more than once.

$f(7) = \mathbf{7^2 + 7}$ Every *x* must be replaced with 7.

$\mathbf{= 49 + 7}$

$\mathbf{= 56}$

(b) $f(x) = x^2 + 5x$

$f(-2) = \mathbf{(-2)^2 + 5 \times -2}$ ☞ Every *x* must be replaced with –2.

$\mathbf{= 4 - 10}$

$\mathbf{= -6}$

(c) $f(x) = 2x^2 - 6x$

$f(5) = \mathbf{2 \times 5^2 - 6 \times 5}$ ☞ Every *x* must be replaced with 5.

$\mathbf{= 2 \times 25 - 30}$

$\mathbf{= 50 - 30}$

$\mathbf{= 20}$

(d) $f(x) = x^2 - x - 1$

$f(3) = \mathbf{3^2 - 3 - 1}$

$\mathbf{= 9 - 3 - 1}$

$\mathbf{= 5}$

(e) $f(x) = 4x - x^2$

$f(-1) = \mathbf{4 \times -1 - (-1)^2}$

$\mathbf{= -4 - 1}$

$\mathbf{= -5}$

(f) $f(x) = x^3 + x^2 + x$

$f(-2) = \mathbf{(-2)^3 + (-2)^2 + (-2)}$

$\mathbf{= -8 + 4 - 2}$

$\mathbf{= -6}$

(g) $f(t) = t^2$

$f(5) = \mathbf{5^2}$

$\mathbf{= 25}$

(h) $g(y) = 5y - 1$

$g(4) = \mathbf{5 \times 4 - 1}$

$\mathbf{= 20 - 1}$

$\mathbf{= 19}$

(i) $h(b) = 3b^2 - b$

$h(3) = \mathbf{3 \times 3^2 - 3}$

$\mathbf{= 3 \times 9 - 3}$

$\mathbf{= 27 - 3}$

$\mathbf{= 24}$

50 EQUATIONS REVIEW

Equations can be solved by:

- adding the same number to both sides of an equation;
- subtracting the same number from both sides of the equation;
- multiplying both sides of the equation by the same number;
- dividing both sides of the equation by the same number.

Example:

Solve:

(a) $x + 8 = 15$ **(c)** $f - 5 = -2$ **(e)** $-5h = -15$ **(g)** $\frac{e}{6} = 12$ **(i)** $\frac{-t}{4} = 4$

(b) $w + 9 = 7$ **(d)** $3x = -12$ **(f)** $-k = 8$ **(h)** $\frac{m}{-2} = -8$

Solution:

(a) $x + 8 = 15$
$x = 15 - 8$
$x = 7$

(b) $w + 9 = 7$
$w = 7 - 9$
$w = -2$

(c) $f - 5 = -2$
$f = -2 + 5$
$f = 3$

(d) $3x = -12$
$x = -12 \div 3$
$x = -4$

(e) $-5h = -15$
$h = -15 \div -5$
$h = 3$

(f) $-k = 8$
$k = 8 \div -1$
$k = -8$

(g) $\frac{e}{6} = 12$
$e = 12 \times 6$
$e = 72$

(h) $\frac{m}{-2} = -8$
$m = -8 \times -2$
$m = 16$

(i) $\frac{-t}{4} = 4$
$t = 4 \times -4$
$t = -16$

Finding the solution to an equation might involve more than one step.
Remember: Do the same thing to both sides of the equation.

Example:

Solve:

(a) $2k - 6 = 35$ **(b)** $3e + 8 = -7$ **(c)** $\frac{t}{2} + 9 = 1$

Solution:

(a) $2k - 6 = 35$
$2k = 35 + 6$
$2k = 41$
$k = 41 \div 2$
$k = 20.5$

(b) $3e + 8 = -7$
$3e = -7 - 8$
$3e = -15$
$e = -15 \div 3$
$e = -5$

(c) $\frac{t}{2} + 9 = 1$
$\frac{t}{2} = 1 - 9$
$\frac{t}{2} = -8$
$t = -8 \times 2$
$t = -16$

We can always check, by substitution, whether we have the right solution to an equation.

For example: $\frac{t}{2} + 9 = 1, \quad t = -16$

$\frac{-16}{2} + 9 = 1$

$-8 + 9 = 1$

$1 = 1$ ✓

1. Solve:

(a) $a + 9 = 13$

(b) $e - 14 = 35$

(c) $t + 4 = 1$

(d) $h - 6 = -7$

(e) $y - 4 = -2$

(f) $k + 9 = -5$

(g) $7d = 84$

(h) $-3m = 18$

(i) $8c = -12$

(j) $-9b = -144$

(k) $-n = 2$

(l) $\frac{x}{5} = 15$

(m) $\frac{z}{6} = -12$

(n) $\frac{-p}{7} = -63$

(o) $\frac{s}{-12} = -12$

(p) $f - 13 = -4$

(q) $-17w = 102$

(r) $15 + g = -13$

(s) $\frac{-q}{18} = 0$

(t) $y + 49 = 20$

2. Solve:

(a) $3n + 4 = 19$

(b) $5t - 81 = 46$

(c) $9v + 17 = -6$

(d) $4f - 6 = -6$

(e) $6g + 13 = -13$

(f) $4m = 30 - m$

(g) $2k - 9 = -15$

(h) $7d = 58 + 9d$

(i) $\frac{c}{8} + 3 = 7$

(j) $\frac{r}{2} - 39 = 41$

(k) $\frac{h}{3} + 15 = 2$

(l) $\frac{b}{7} - 18 = -20$

3. Show, by substitution, that $n = 13$ is the solution to the equation $39 - 9n = -6n$.

52 EQUATIONS REVIEW Worked Solutions

1. Solve:

(a) $a + 9 = 13$

$\mathbf{a = 13 - 9}$ ☞ Subtract 9 from both sides of the equation.

$\mathbf{a = 4}$

(b) $e - 14 = 35$

$\mathbf{e = 35 + 14}$ ☞ Add 14 to both sides of the equation.

$\mathbf{e = 49}$

(c) $t + 4 = 1$

$\mathbf{t = 1 - 4}$

$\mathbf{t = -3}$

(d) $h - 6 = -7$

$\mathbf{h = -7 + 6}$

$\mathbf{h = -1}$

(e) $y - 4 = -2$

$\mathbf{y = -2 + 4}$

$\mathbf{y = 2}$

(f) $k + 9 = -5$

$\mathbf{k = -5 - 9}$

$\mathbf{k = -14}$

(g) $7d = 84$

$\mathbf{d = 84 \div 7}$ ☞ Divide both sides of the equation by 7.

$\mathbf{d = 12}$

(h) $-3m = 18$

$\mathbf{m = 18 \div -3}$

$\mathbf{m = -6}$

(i) $8c = -12$

$\mathbf{c = -12 \div 8}$

$\mathbf{c = -1.5}$

(j) $-9b = -144$

$\mathbf{b = -144 \div -9}$

$\mathbf{b = 16}$

(k) $-n = 2$

$\mathbf{n = 2 \div -1}$ ☞ $-n$ means $-1n$.
We divide both sides by –1.

$\mathbf{n = -2}$

(l) $\frac{x}{5} = 15$

$\mathbf{x = 15 \times 5}$ ☞ We multiply both sides of the equation by 5.

$\mathbf{x = 75}$

(m) $\frac{z}{6} = -12$

$\mathbf{z = -12 \times 6}$

$\mathbf{z = -72}$

(n) $\frac{-p}{7} = -63$

$\mathbf{p = -63 \times -7}$

$\mathbf{p = 441}$

(o) $\frac{s}{-12} = -12$

$\mathbf{s = -12 \times -12}$

$\mathbf{s = 144}$

(p) $f - 13 = -4$

$\mathbf{f = -4 + 13}$

$\mathbf{f = 9}$

(q) $-17w = 102$

$\mathbf{w = 102 \div -17}$

$\mathbf{w = -6}$

(r) $15 + g = -13$

$\mathbf{g = -13 - 15}$

$\mathbf{g = -28}$

(s) $\frac{-q}{18} = 0$

$\mathbf{q = 0 \times -18}$

$\mathbf{q = 0}$

(t) $y + 49 = 20$

$\mathbf{y = 20 - 49}$

$\mathbf{y = -29}$

2. Solve:

(a) $3n + 4 = 19$

$\mathbf{3n = 19 - 4}$ ☞ First we subtract 4 from both sides of the equation,

$\mathbf{3n = 15}$

$\mathbf{n = 15 \div 3}$ ☞ then divide both sides of the equation by 3.

$\mathbf{n = 5}$

(b) $5t - 81 = 46$

$\mathbf{5t = 46 + 81}$ ☞ First we add 81 to both sides of the equation,

$\mathbf{5t = 127}$

$\mathbf{t = 127 \div 5}$ ☞ then divide both sides of the equation by 5.

$\mathbf{t = 25.4}$

(c) $9v + 17 = -6$

$\mathbf{9v = -6 - 17}$

$\mathbf{9v = -23}$

$\mathbf{v = -23 \div 9}$

$\mathbf{v = -2\frac{5}{9}}$

(d) $4f - 6 = -6$

$\mathbf{4f = -6 + 6}$

$\mathbf{4f = 0}$

$\mathbf{f = 0 \div 4}$

$\mathbf{f = 0}$

(e) $6g + 13 = -13$

$\mathbf{6g = -13 - 13}$

$\mathbf{6g = -26}$

$\mathbf{g = -26 \div 6}$

$\mathbf{g = -4\frac{1}{3}}$

(f) $4m = 30 - m$

$\mathbf{4m + m = 30}$ ☞ Add m to both sides of the equation.

$\mathbf{5m = 30}$

$\mathbf{m = 30 \div 5}$

$\mathbf{m = 6}$

(g) $2k - 9 = -15$

$\mathbf{2k = -15 + 9}$

$\mathbf{2k = -6}$

$\mathbf{k = -6 \div 2}$

$\mathbf{k = -3}$

(h) $7d = 58 + 9d$

$\mathbf{7d - 9d = 58}$

$\mathbf{-2d = 58}$

$\mathbf{d = 58 \div -2}$

$\mathbf{d = -29}$

(i) $\frac{c}{8} + 3 = 7$

$\mathbf{\frac{c}{8} = 7 - 3}$ ☞ First subtract 3 from both sides of the equation,

$\mathbf{\frac{c}{8} = 4}$

$\mathbf{c = 4 \times 8}$ ☞ then multiply both sides of the equation by 8.

$\mathbf{c = 32}$

(j) $\frac{r}{2} - 39 = 41$

$\mathbf{\frac{r}{2} = 41 + 39}$

$\mathbf{\frac{r}{2} = 80}$

$\mathbf{r = 80 \times 2}$

$\mathbf{r = 160}$

(k) $\frac{h}{3} + 15 = 2$

$\mathbf{\frac{h}{3} = 2 - 15}$

$\mathbf{\frac{h}{3} = -13}$

$\mathbf{h = -13 \times 3}$

$\mathbf{h = -39}$

(l) $\frac{b}{7} - 18 = -20$

$\mathbf{\frac{b}{7} = -20 + 18}$

$\mathbf{\frac{b}{7} = -2}$

$\mathbf{b = -2 \times 7}$

$\mathbf{b = -14}$

3. Show, by substitution, that $n = 13$ is the solution to the equation $39 - 9n = -6n$.

$\mathbf{39 - 9n = -6n}$ ☞ First write the equation,

When $\mathbf{n = 13}$**,** $\mathbf{39 - 9 \times 13 = -6 \times 13}$ ☞ then replace every n with 13.

$\mathbf{39 - 117 = -78}$

$\mathbf{-78 = -78}$ ✓ ☞ Is this true or false?

$\therefore$ $\mathbf{n = 13}$ **is the solution to the equation** $\mathbf{39 - 9n = -6n}$**.**

54 FURTHER EQUATIONS

If there are pronumerals as well as numbers on both sides of an equation, we need to get the pronumerals onto one side and the numbers onto the other in order to solve the equation.

It doesn't matter whether we remove the pronumerals first, the numbers first, or do both at the same time, as long as we do the same thing to both sides of our equation.

Example:

Solve:

(a) $4x - 10 = 2x + 18$ **(b)** $7e + 9 = 35 - 3e$ **(c)** $8t + 17 = 3t - 13$ **(d)** $9f - 20 = 4f - 70$

Solution:

(a)
$$4x - 10 = 2x + 18$$
$$4x = 2x + 18 + 10$$
$$4x = 2x + 28$$
$$4x - 2x = 28$$
$$2x = 28$$
$$x = 14$$

(b)
$$7e + 9 = 35 - 3e$$
$$7e + 3e + 9 = 35$$
$$10e + 9 = 35$$
$$10e = 35 - 9$$
$$10e = 26$$
$$e = 2.6$$

(c)
$$8t + 17 = 3t - 13$$
$$8t = 3t - 13 - 17$$
$$8t = 3t - 30$$
$$8t - 3t = -30$$
$$5t = -30$$
$$t = -6$$

(d)
$$9f - 20 = 4f - 70$$
$$9f = 4f - 70 + 20$$
$$9f = 4f - 50$$
$$9f - 4f = -50$$
$$5f = -50$$
$$f = -10$$

We can use our knowledge of these types of equations to solve problems.

Example:

Find the value of a.

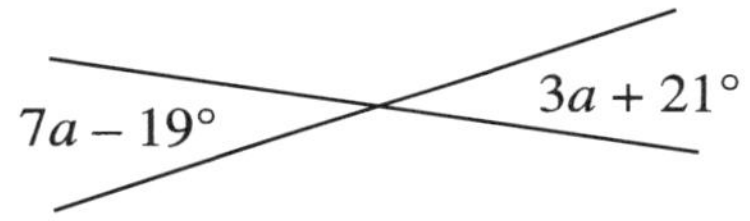

Solution:

$7a - 19 = 3a + 21$ (Vertically opposite angles are equal.)

$7a = 3a + 40$ (Adding 19 to both sides.)

$4a = 40$ (Subtracting $3a$ from both sides.)

$a = 10$ (Dividing both sides by 4.)

Example:

5 less than 3 times a number is equal to 7 more than twice the number. Find the number.

Solution:

Let the number be x.

$3x - 5 = 2x + 7$

$3x = 2x + 12$ (Adding 5 to both sides.)

$x = 12$ (Subtracting $2x$ from both sides.)

$\therefore$ **the number is 12.**

1. Solve:

(a) $9c + 4 = 5c + 34$

(b) $8p - 22 = 3p + 56$

(c) $5e + 38 = 2e + 5$

(d) $6t - 37 = -3t + 8$

(e) $7u + 85 = 13 - 2u$

(f) $4m - 19 = 37 - 4m$

(g) $11a + 15 = 8a - 19$

(h) $9k - 14 = 8k - 32$

(i) $12q - 17 = -19 - 2q$

2. Opposite angles of a parallelogram are equal. Find the value of a.

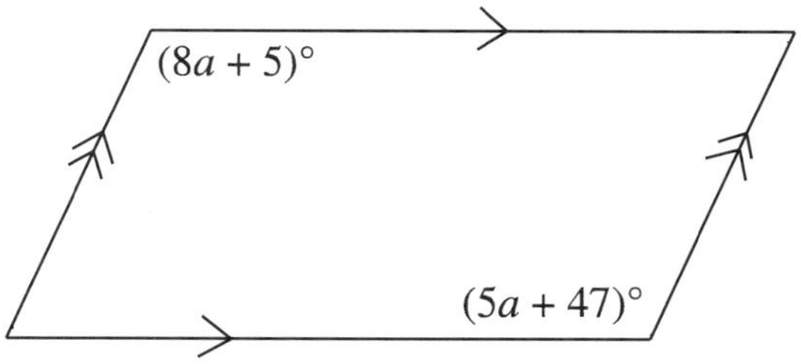

3. Nine less than five times a certain number is equal to 17 more than three times the number. Find the number.

4. When 15 is subtracted from four times a certain number, the result is 7 more than three times the number. What is the number?

56 FURTHER EQUATIONS Worked Solutions

1. Solve:

(a) $9c + 4 = 5c + 34$

$9c = 5c + 34 - 4$ ☞ Subtract 4 from both sides of the equation.

$9c = 5c + 30$

$9c - 5c = 30$ Subtract $5c$ from both sides of the equation.

$4c = 30$

$c = 30 \div 4$ Divide both sides of the equation by 4.

$c = 7.5$

(b) $8p - 22 = 3p + 56$

$8p = 3p + 56 + 22$ ☞ Add 22 to both sides of the equation.

$8p = 3p + 78$

$8p - 3p = 78$ Subtract $3p$ from both sides of the equation.

$5p = 78$

$p = 78 \div 5$ Divide both sides of the equation by 5.

$p = 15.6$

(c) $5e + 38 = 2e + 5$

$5e = 2e + 5 - 38$ ☞ We could subtract $2e$ from both sides first, then subtract 38, or we could do both at the same time.

$5e = 2e - 33$

$5e - 2e = -33$

$3e = -33$

$e = -33 \div 3$

$e = -11$

(d) $6t - 37 = -3t + 8$

$6t = -3t + 8 + 37$ ☞ Add 37 to both sides of the equation.

$6t = -3t + 45$

$6t + 3t = 45$ Add $3t$ to both sides of the equation.

$9t = 45$

$t = 45 \div 9$ Divide both sides of the equation by 9.

$t = 5$

(e) $7u + 85 = 13 - 2u$

$7u = 13 - 2u - 85$

$7u = -2u - 72$

$7u + 2u = -72$

$9u = -72$

$u = -72 \div 9$

$u = -8$

(f) $4m - 19 = 37 - 4m$

$4m = 37 - 4m + 19$

$4m = 56 - 4m$

$4m + 4m = 56$

$8m = 56$

$m = 56 \div 8$

$m = 7$

(g) $11a + 15 = 8a - 19$
$\mathbf{11a = 8a - 19 - 15}$
$\mathbf{11a = 8a - 34}$
$\mathbf{11a - 8a = -34}$
$\mathbf{3a = -34}$
$\mathbf{a = -34 \div 3}$
$\mathbf{a = -11\frac{1}{3}}$

(h) $9k - 14 = 8k - 32$
$\mathbf{9k = 8k - 32 + 14}$
$\mathbf{9k = 8k - 18}$
$\mathbf{9k - 8k = -18}$
$\mathbf{k = -18}$

(i) $12q - 17 = -19 - 2q$
$\mathbf{12q = -19 - 2q + 17}$
$\mathbf{12q = -2 - 2q}$
$\mathbf{12q + 2q = -2}$
$\mathbf{14q = -2}$
$\mathbf{q = -2 \div 14}$
$\mathbf{q = -\frac{1}{7}}$

2. Opposite angles of a parallelogram are equal. Find the value of a.

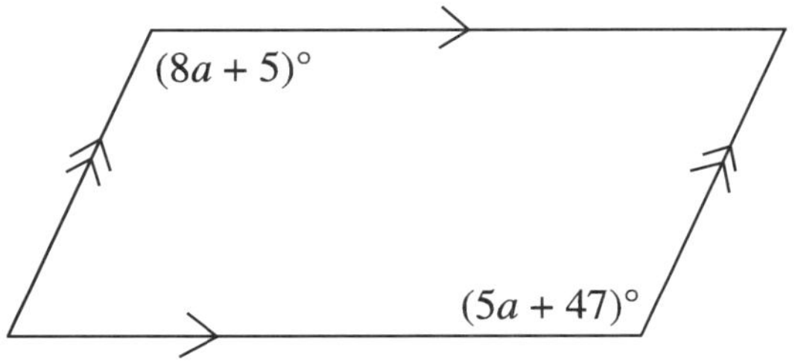

$\mathbf{8a + 5 = 5a + 47}$ ☞ Opposite angles of a parallelogram are equal.
$\mathbf{8a = 5a + 47 - 5}$
$\mathbf{8a = 5a + 42}$
$\mathbf{8a - 5a = 42}$
$\mathbf{3a = 42}$
$\mathbf{a = 42 \div 3}$
$\therefore$ $\mathbf{a = 14}$ (And the size of each of those angles is 117°.)

3. Nine less than five times a certain number is equal to 17 more than three times the number. Find the number.

Let the number be *n*. ☞ Begin with a statement like this. (You can use any pronumeral.)

$\mathbf{5n - 9 = 3n + 17}$ Write the statement using symbols.
$\mathbf{5n = 3n + 17 + 9}$ Solve the equation.
$\mathbf{5n = 3n + 26}$
$\mathbf{5n - 3n = 26}$
$\mathbf{2n = 26}$
$\mathbf{n = 26 \div 2}$
$\mathbf{n = 13}$ (Check that 13 is the number!)
$\therefore$ **the number is 13.** Write the answer in a sentence.

4. When 15 is subtracted from four times a certain number, the result is 7 more than three times the number. What is the number?

Let the number be *x*. ☞ Write a statement like this first.

$\mathbf{4x - 15 = 3x + 7}$
$\mathbf{4x = 3x + 22}$ Adding 15 to both sides of the equation.
$\mathbf{4x - 3x = 22}$
$\mathbf{x = 22}$
$\therefore$ **the number is 22.** Write the answer in a sentence.

EQUATIONS WITH GROUPING SYMBOLS

In solving an equation, we might need to remove grouping symbols.

Example:

Solve:

(a) $3(x + 8) = 27$

(b) $5(2x - 7) = 2(4x + 1)$

Solution:

(a) $3(x + 8) = 27$

$3x + 24 = 27$

$3x = 27 - 24$

$3x = 3$

$x = 3 \div 3$

$x = 1$

(b) $5(2x - 7) = 2(4x + 1)$

$10x - 35 = 8x + 2$

$10x = 8x + 2 + 35$

$10x = 8x + 37$

$10x - 8x = 37$

$2x = 37$

$x = 37 \div 2$

$x = 18.5$

Before attempting to actually solve an equation, we should collect any like terms.

Example:

Solve:

(a) $7k = 54 - 4(k - 3)$

(b) $4(2t - 3) - (5t + 2) = 9$

(c) $7(x + 3) - 4(x - 1) = 0$

(d) $8(3y + 4) - 5 = 17 - 2(6y - 7)$

Solution:

(a) $7k = 54 - 4(k - 3)$

$7k = 54 - 4k + 12$

$7k = 66 - 4k$

$7k + 4k = 66$

$11k = 66$

$k = 66 \div 11$

$k = 6$

(b) $4(2t - 3) - (5t + 2) = 9$

$8t - 12 - 5t - 2 = 9$

$3t - 14 = 9$

$3t = 9 + 14$

$3t = 23$

$t = 23 \div 3$

$t = 7\frac{2}{3}$

(c) $7(x + 3) - 4(x - 1) = 0$

$7x + 21 - 4x + 4 = 0$

$3x + 25 = 0$

$3x = -25$

$x = -25 \div 3$

$x = -8\frac{1}{3}$

(d) $8(3y + 4) - 5 = 17 - 2(6y - 7)$

$24y + 32 - 5 = 17 - 12y + 14$

$24y + 27 = 31 - 12y$

$36y + 27 = 31$

$36y = 4$

$y = \frac{4}{36}$ $(4 \div 36)$

1

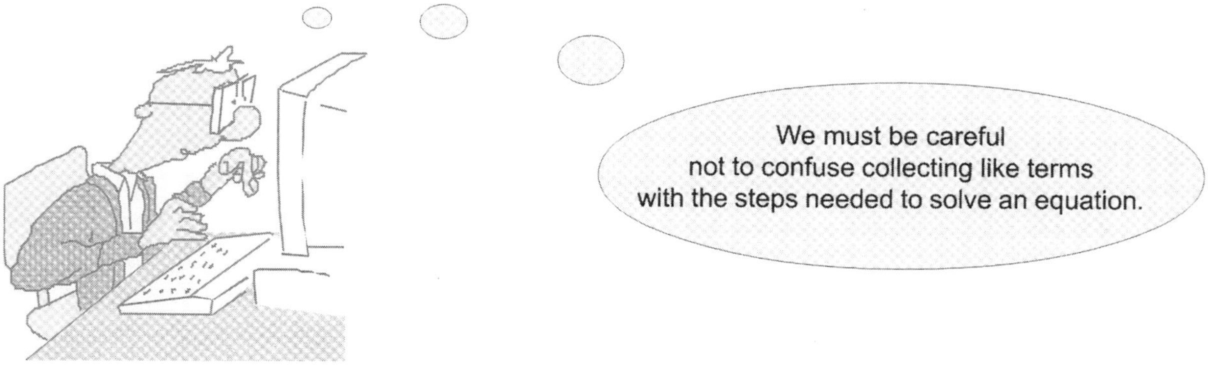

1. **Solve:**
   (a) $4(k + 7) = 64$
   (b) $6(5e - 9) = 84$
   (c) $7(8y - 1) = -35$
   (d) $9(3x + 4) = 8(x - 5)$
   (e) $7(7a - 2) = 5(2a + 5)$
   (f) $12(5d - 13) = 3(10 - 11d)$

2. **Solve:**
   (a) $12p = 34 - 2(8p + 3)$
   (b) $6(5w + 1) + 1 = 73$
   (c) $4(3t - 5) + 7 = 5t - 3$
   (d) $11(3f - 4) = 7(5f - 8)$
   (e) $8(4c + 7) = 5(3c + 1)$
   (f) $9(4h - 1) = 5(7h - 1)$
   (g) $3(2g + 9) + 8 = 2(2g - 9) - 11$
   (h) $7(3q + 14) - 31 = 76 - 6(2q + 7)$

EQUATIONS WITH GROUPING SYMBOLS Worked Solutions

1. Solve:

(a) $4(k + 7) = 64$

$\mathbf{4k + 28 = 64}$ ☞ First remove the grouping symbols,

$\mathbf{4k = 64 - 28}$ then subtract 28 from both sides of the equation.

$\mathbf{4k = 36}$

$\mathbf{k = 36 \div 4}$ ☞ Divide both sides of the equation by 4.

$\mathbf{k = 9}$

(b) $6(5e - 9) = 84$

$\mathbf{30e - 54 = 84}$

$\mathbf{30e = 84 + 54}$

$\mathbf{30e = 138}$

$\mathbf{e = 138 \div 30}$

$\mathbf{e = 4.6}$

(c) $7(8y - 1) = -35$

$\mathbf{56y - 7 = -35}$

$\mathbf{56y = -35 + 7}$

$\mathbf{56y = -28}$

$\mathbf{y = -28 \div 56}$

$\mathbf{y = -0.5}$

(d) $9(3x + 4) = 8(x - 5)$

$\mathbf{27x + 36 = 8x - 40}$ ☞ Remove both sets of brackets,

$\mathbf{27x = 8x - 40 - 36}$ then subtract 36 from both sides.

$\mathbf{27x = 8x - 76}$

$\mathbf{27x - 8x = -76}$ ☞ Subtract $8x$ from both sides.

$\mathbf{19x = -76}$

$\mathbf{x = -76 \div 19}$ ☞ Divide both sides of the equation by 19.

$\mathbf{x = -4}$

(e) $7(7a - 2) = 5(2a + 5)$

$\mathbf{49a - 14 = 10a + 25}$

$\mathbf{49a = 10a + 25 + 14}$

$\mathbf{49a = 10a + 39}$

$\mathbf{49a - 10a = 39}$

$\mathbf{39a = 39}$

$\mathbf{a = 39 \div 39}$

$\mathbf{a = 1}$

(f) $12(5d - 13) = 3(10 - 11d)$

$\mathbf{60d - 156 = 30 - 33d}$

$\mathbf{60d = 30 - 33d + 156}$

$\mathbf{60d = 186 - 33d}$

$\mathbf{60d + 33d = 186}$

$\mathbf{93d = 186}$

$\mathbf{d = 186 \div 93}$

$\mathbf{d = 2}$

2. Solve:

(a) $12p = 34 - 2(8p + 3)$ ☞ First remove the brackets, being very careful

$\mathbf{12p = 34 - 16p - 6}$ to get the signs correct,

$\mathbf{12p = 28 - 16p}$ then collect the like terms.

$\mathbf{12p + 16p = 28}$ ☞ Add $16p$ to both sides of the equation.

$\mathbf{28p = 28}$

$\mathbf{p = 28 \div 28}$ ☞ Divide both sides of the equation by 28.

$\mathbf{p = 1}$

(b) $6(5w + 1) + 1 = 73$

$30w + 6 + 1 = 73$

$30w + 7 = 73$

$30w = 73 - 7$

$30w = 66$

$w = 66 \div 30$

$w = 2.2$

(c) $4(3t - 5) + 7 = 5t - 3$

$12t - 20 + 7 = 5t - 3$

$12t - 13 = 5t - 3$

$12t = 5t - 3 + 13$

$12t = 5t + 10$

$12t - 5t = 10$

$7t = 10$

$t = 10 \div 7$

$t = 1\frac{3}{7}$

(d) $11(3f - 4) = 7(5f - 8)$

$33f - 44 = 35f - 56$

$33f = 35f - 56 + 44$

$33f = 35f - 12$

$33f - 35f = -12$

$-2f = -12$

$f = -12 \div -2$

$f = 6$

(e) $8(4c + 7) = 5(3c + 1)$

$32c + 56 = 15c + 5$

$32c = 15c + 5 - 56$

$32c = 15c - 51$

$32c - 15c = -51$

$17c = -51$

$c = -51 \div 17$

$c = -3$

(f) $9(4h - 1) = 5(7h - 1)$

$36h - 9 = 35h - 5$

$36h = 35h - 5 + 9$

$36h = 35h + 4$

$36h - 35h = 4$

$h = 4$

(g) $3(2g + 9) + 8 = 2(2g - 9) - 11$

$6g + 27 + 8 = 4g - 18 - 11$ — First remove all brackets, then collect like terms.

$6g + 35 = 4g - 29$

$6g = 4g - 29 - 35$

$6g = 4g - 64$

$6g - 4g = -64$

$2g = -64$

$g = -64 \div 2$

$g = -32$

(h) $7(3q + 14) - 31 = 76 - 6(2q + 7)$

$21q + 98 - 31 = 76 - 12q - 42$ — Remember, be careful with the signs.

$21q + 67 = 34 - 12q$

$21q = 34 - 12q - 67$ — Subtract 67 from both sides of the equation.

$21q = -33 - 12q$

$21q + 12q = -33$ — Add $12q$ to both sides.

$33q = -33$

$q = -33 \div 33$ — Divide both sides by 33.

$q = -1$

62 EQUATIONS INVOLVING FRACTIONS

We know that equations like $\frac{x}{2} = 6$ can be solved by multiplying both sides of the equation.

Example:

Solve:

(a) $\frac{2x}{3} = 11$ **(b)** $\frac{4k-1}{3} = -2$

Solution:

(a) $\frac{2x}{3} = 11$

$2x = 11 \times 3$

$2x = 33$

$x = 33 \div 2$

$x = 16\frac{1}{2}$

(b) $\frac{4k-1}{3} = -2$

$4k - 1 = -2 \times 3$

$4k - 1 = -6$

$4k = -5$ (adding 1)

$k = -1.25$ (dividing by 4)

Equations involving fractions are solved by multiplying both sides of the equation by a common denominator.

Example:

Solve:

(a) $\frac{2h}{5} = \frac{1}{3}$ **(b)** $\frac{5a+6}{4} = \frac{3a+1}{5}$

Every term in the equation must be multiplied.

Solution:

(a) $\frac{2h}{5} = \frac{1}{3}$

$\overset{3}{\cancel{15}} \times \frac{2h}{\cancel{5}_1} = \overset{5}{\cancel{15}} \times \frac{1}{\cancel{3}_1}$

$3 \times 2h = 5 \times 1$

$6h = 5$

$h = \frac{5}{6}$

(b) $\frac{5a+6}{4} = \frac{3a+1}{5}$

$\overset{5}{\cancel{20}} \times \left(\frac{5a+6}{\cancel{4}_1}\right) = \overset{4}{\cancel{20}} \times \left(\frac{3a+1}{\cancel{5}_1}\right)$

$5(5a + 6) = 4(3a + 1)$

$25a + 30 = 12a + 4$

$13a = -26$

$a = -2$

We use brackets to help keep track of what must be multiplied.

Example:

Solve:

(a) $\frac{3x}{8} + 9 = \frac{x-1}{6}$ **(b)** $\frac{t+4}{9} + \frac{2t}{5} = 3$ **(c)** $\frac{4m+1}{4} - \frac{m-8}{2} = \frac{3m-1}{8}$

Solution:

(a) $\frac{3x}{8} + 9 = \frac{x-1}{6}$

$\underset{3}{\cancel{24}} \times \frac{3x}{{}_1\cancel{8}} + 24 \times 9 = \underset{4}{\cancel{24}} \times \left(\frac{x-1}{\cancel{6}_1}\right)$

$9x + 216 = 4(x - 1)$

$9x + 216 = 4x - 4$

$5x = -220$

$x = -44$

(b) $\frac{t+4}{9} + \frac{2t}{5} = 3$

$\underset{5}{\cancel{45}} \times \left(\frac{t+4}{\cancel{9}_1}\right) + \underset{9}{\cancel{45}} \times \frac{2t}{\cancel{5}_1} = 45 \times 3$

$5(t + 4) + 18t = 135$

$5t + 20 + 18t = 135$

$23t + 20 = 135$

$23t = 115$

$t = 5$

(c) $\frac{4m+1}{4} - \frac{m-8}{2} = \frac{3m-1}{8}$

$\underset{2}{\cancel{8}} \times \left(\frac{4m+1}{\cancel{4}_1}\right) - \underset{4}{\cancel{8}} \times \left(\frac{m-8}{\cancel{2}_1}\right) = \underset{1}{\cancel{8}} \times \left(\frac{3m-1}{\cancel{8}_1}\right)$

$2(4m + 1) - 4(m - 8) = 3m - 1$

$8m + 2 - 4m + 32 = 3m - 1$

$4m + 34 = 3m - 1$

$m + 34 = -1$

$m = -35$

1. **Solve:**

(a) $\frac{5a}{9} = 12$

(b) $\frac{3x}{8} = \frac{1}{2}$

(c) $\frac{10p}{3} = -4$

(d) $-\frac{2k}{7} = 8$

(e) $\frac{3h - 1}{4} = 5$

(f) $\frac{5b + 6}{7} = -2$

(g) $\frac{2e}{9} = \frac{3}{5}$

(h) $\frac{5y}{6} = -\frac{7}{8}$

2. **Solve:**

(a) $\frac{6t + 5}{7} = \frac{2t + 13}{3}$

(b) $\frac{5g - 4}{6} = \frac{9 - 2g}{5}$

(c) $\frac{8e + 3}{7} = \frac{e - 3}{2}$

3. **Solve: (making sure to multiply *every* term by the common denominator)**

(a) $\frac{2m}{5} + 7 = \frac{m + 2}{3}$

(b) $\frac{2q + 13}{7} + \frac{3q + 2}{4} = 21$

(c) $\frac{4u - 5}{6} + 3u - 4 = \frac{13u - 4}{2}$

(d) $\frac{10m + 3}{4} - \frac{2m - 1}{5} = \frac{5m + 10}{3}$

64 EQUATIONS INVOLVING FRACTIONS Worked Solutions

1. Solve:

(a) $\frac{5a}{9} = 12$

$\mathbf{5a = 12 \times 9}$

Multiply both sides of the equation by 9.

$\mathbf{5a = 108}$

$\mathbf{a = 108 \div 5}$

$\mathbf{a = 21.6}$

(b) $\frac{3x}{8} = \frac{1}{2}$

$\mathbf{3x = \frac{1}{2} \times 8}$

$\mathbf{3x = 4}$

$\mathbf{x = 4 \div 3}$

$\mathbf{x = 1\frac{1}{3}}$

(c) $\frac{10p}{3} = -4$

$\mathbf{10p = -4 \times 3}$

$\mathbf{10p = -12}$

$\mathbf{p = -12 \div 10}$

$\mathbf{p = -1.2}$

(d) $-\frac{2k}{7} = 8$

$\mathbf{-2k = 8 \times 7}$

$\mathbf{-2k = 56}$

$\mathbf{k = 56 \div -2}$

$\mathbf{k = -28}$

(e) $\frac{3h - 1}{4} = 5$

$\mathbf{3h - 1 = 5 \times 4}$

Multiply both sides of the equation by 4.

$\mathbf{3h - 1 = 20}$

$\mathbf{3h = 21}$

Adding 1 to both sides.

$\mathbf{h = 7}$

Dividing both sides by 3.

(f) $\frac{5b + 6}{7} = -2$

$\mathbf{5b + 6 = -2 \times 7}$

Multiply both sides of the equation by 7.

$\mathbf{5b + 6 = -14}$

$\mathbf{5b = -20}$

Subtracting 6 from both sides.

$\mathbf{b = -4}$

Dividing both sides by 5.

(g) $\frac{2e}{9} = \frac{3}{5}$

The common denominator is 45.

$\overset{5}{\cancel{45}} \times \frac{2e}{\cancel{9}_1} = \overset{9}{\cancel{45}} \times \frac{3}{\cancel{5}_1}$

Multiply both sides of the equation by 45.

$\mathbf{10e = 27}$

$\mathbf{e = 2.7}$

Dividing both sides by 10.

(h) $\frac{5y}{6} = -\frac{7}{8}$

The lowest common denominator is 24.

$\overset{4}{\cancel{24}} \times \frac{5y}{\cancel{6}_1} = \overset{3}{\cancel{24}} \times \frac{-7}{\cancel{8}_1}$

Multiply both sides of the equation by 24.

$\mathbf{20y = -21}$

$\mathbf{y = -1.05}$

Dividing both sides of the equation by 20.

2. Solve:

(a) $\frac{6t + 5}{7} = \frac{2t + 13}{3}$

The lowest common denominator is 21.

$\overset{3}{\cancel{21}} \times \left(\frac{6t + 5}{\cancel{7}_1}\right) = \overset{7}{\cancel{21}} \times \left(\frac{2t + 13}{\cancel{3}_1}\right)$

Multiply both sides of the equation by 21.

$\mathbf{3(6t + 5) = 7(2t + 13)}$

Use brackets (all the numerator must be multiplied).

$\mathbf{18t + 15 = 14t + 91}$

$\mathbf{18t = 14t + 76}$

Subtracting 15 from both sides of the equation.

$\mathbf{4t = 76}$

Subtracting 14t from both sides.

$\mathbf{t = 19}$

Dividing both sides of the equation by 4.

(b) $\frac{5g-4}{6} = \frac{9-2g}{5}$

$$^{5}\cancel{30} \times \left(\frac{5g-4}{\cancel{6}_1}\right) = {}^{6}\cancel{30} \times \left(\frac{9-2g}{\cancel{5}_1}\right)$$
$$5(5g-4) = 6(9-2g)$$
$$25g - 20 = 54 - 12g$$
$$25g = 74 - 12g$$
$$37g = 74$$
$$g = 2$$

(c) $\frac{8e+3}{7} = \frac{e-3}{2}$

$$^{2}\cancel{14} \times \left(\frac{8e+3}{\cancel{7}_1}\right) = {}^{7}\cancel{14} \times \left(\frac{e-3}{\cancel{2}_1}\right)$$
$$2(8e+3) = 7(e-3)$$
$$16e + 6 = 7e - 21$$
$$16e = 7e - 27$$
$$9e = -27$$
$$e = -3$$

3. Solve: (making sure to multiply *every* term by the common denominator)

(a) $\frac{2m}{5} + 7 = \frac{m+2}{3}$

$$^{3}\cancel{15} \times \frac{2m}{\cancel{5}_1} + 15 \times 7 = {}^{5}\cancel{15} \times \left(\frac{m+2}{\cancel{3}_1}\right)$$
$$6m + 105 = 5(m+2)$$
$$6m + 105 = 5m + 10$$
$$6m = 5m - 95$$
$$m = -95$$

☞ Multiply every term by the common denominator.

Use a bracket.

Subtract 105 from both sides of the equation.
Subtract 5m from both sides.

(b) $\frac{2q+13}{7} + \frac{3q+2}{4} = 21$

$$_{4}\cancel{28} \times \left(\frac{2q+13}{\cancel{7}_1}\right) + {}_{7}\cancel{28} \times \left(\frac{3q+2}{\cancel{4}_1}\right) = 28 \times 21$$
$$4(2q+13) + 7(3q+2) = 588$$
$$8q + 52 + 21q + 14 = 588$$
$$29q + 66 = 588$$
$$29q = 522$$
$$q = 18$$

(c) $\frac{4u-5}{6} + 3u - 4 = \frac{13u-4}{2}$

$$^{1}\cancel{6} \times \left(\frac{4u-5}{\cancel{6}_1}\right) + 6 \times 3u - 6 \times 4 = {}^{3}\cancel{6} \times \left(\frac{13u-4}{\cancel{2}_1}\right)$$
$$4u - 5 + 18u - 24 = 3(13u - 4)$$
$$22u - 29 = 39u - 12$$
$$22u = 39u + 17$$
$$-17u = 17$$
$$u = -1$$

(d) $\frac{10m+3}{4} - \frac{2m-1}{5} = \frac{5m+10}{3}$

☞ The common denominator is 60.

$$_{15}\cancel{60} \times \left(\frac{10m+3}{{}_1\cancel{4}}\right) - {}_{12}\cancel{60} \times \left(\frac{2m-1}{{}_1\cancel{5}}\right) = {}_{20}\cancel{60} \times \left(\frac{5m+10}{{}_1\cancel{3}}\right)$$

Multiply both sides of the equation by 60.

$$15(10m+3) - 12(2m-1) = 20(5m+10)$$

Use brackets.

$$150m + 45 - 24m + 12 = 100m + 200$$

Remove the brackets, taking care with the signs.

$$126m + 57 = 100m + 200$$

Collect like terms.

$$126m = 100m + 143$$

Subtract 57 from both sides of the equation.

$$26m = 143$$

Subtract 100m from both sides.

$$m = 5.5$$

Divide both sides of the equation by 26.

66 INEQUATIONS

An inequation is an expression involving an inequality sign. Both sides of an inequation are **not** equal.

There are many inequality signs.
Some include:

- $>$ is greater than
- $<$ is less than
- $\geq$ is greater than or equal to
- $\leq$ is less than or equal to
- $\ngtr$ is not greater than
- $\nless$ is not less than
- $\neq$ is not equal to

Inequations are solved like equations.
We must do the same thing to both sides of the inequation.

Example:

Solve:

(a) $x + 9 \geq 11$ **(b)** $7a < 84$ **(c)** $e - 3 \neq 4$ **(d)** $\frac{k}{5} > 10$

Solution:

(a) $x + 9 \geq 11$
$\mathbf{x \geq 11 - 9}$
$\mathbf{x \geq 2}$

(b) $7a < 84$
$\mathbf{a < 84 \div 7}$
$\mathbf{a < 12}$

(c) $e - 3 \neq 4$
$\mathbf{e \neq 4 + 3}$
$\mathbf{e \neq 7}$

(d) $\frac{k}{5} > 10$
$\mathbf{k > 10 \times 5}$
$\mathbf{k > 50}$

To graph on a number line, place a circle around the position of the value, (e.g. if $x > 3$, circle 3).
The circle is open if the value is **not** included ($x > 3$), or solid if the value **is** included ($x \geq 3$).
A line is then drawn in the required direction.

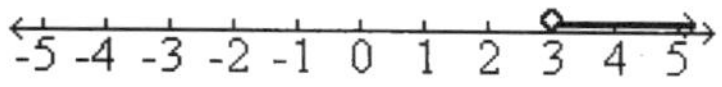

$x > 3$

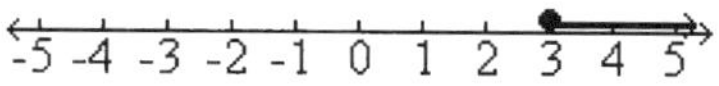

$x \geq 3$

Example:

Graph on a number line:

(a) $x \geq 2$ **(b)** $x < -1$

Solution:

(a) 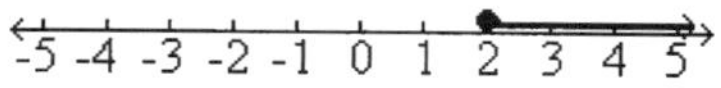

$x \geq 2$

(b) 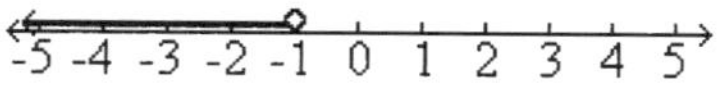

$x < -1$

Solutions to inequations can be graphed on a number line.

Example:

Solve and graph on a number line:

(a) $2x + 5 < 9$ **(b)** $3x - 4 \geq -1$

Solution:

(a) $2x + 5 < 9$
$\mathbf{2x < 4}$
$\mathbf{x < 2}$

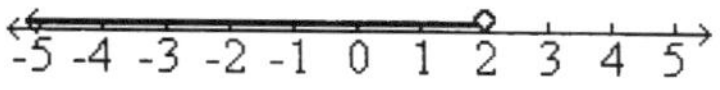

(b) $3x - 4 \geq -1$
$\mathbf{3x \geq 3}$
$\mathbf{x \geq 1}$

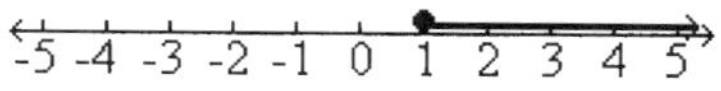

1. **Solve:**

(a) $c + 7 > 13$

(b) $8t < 44$

(c) $k - 17 \leq -6$

(d) $\frac{p}{4} \geq 2$

(e) $6s \neq 108$

(f) $16 + t < 7$

(g) $\frac{a}{5} > -9$

(h) $u - 5 < -19$

(i) $3m + 5 \leq 32$

(j) $2k - 7 \geq 18$

(k) $5b - 3 \neq -7$

(l) $9y + 4 \leq -5$

2. **Graph on the number line provided:**

(a) $x \geq -4$ -5 -4 -3 -2 -1 0 1 2 3 4 5

(b) $x < 1$ -5 -4 -3 -2 -1 0 1 2 3 4 5

(c) $x > -1$ -5 -4 -3 -2 -1 0 1 2 3 4 5

(d) $x \leq 3$ -5 -4 -3 -2 -1 0 1 2 3 4 5

(e) $x < -2$ -5 -4 -3 -2 -1 0 1 2 3 4 5

(f) $x \geq 0$ -5 -4 -3 -2 -1 0 1 2 3 4 5

3. **Solve and graph the solution on the number line provided:**

(a) $2x + 1 \leq 9$

-5 -4 -3 -2 -1 0 1 2 3 4 5

(b) $3x - 7 > -1$

-5 -4 -3 -2 -1 0 1 2 3 4 5

(c) $4x - 4 > -4$

-5 -4 -3 -2 -1 0 1 2 3 4 5

(d) $6x \geq 4x + 8$

-5 -4 -3 -2 -1 0 1 2 3 4 5

INEQUATIONS Worked Solutions

1. Solve:

(a) $c + 7 > 13$

$\mathbf{c > 13 - 7}$ — We solve an inequation like an equation. Subtract 7 from both sides of the inequation.

$\mathbf{c > 6}$

(b) $8t < 44$

$\mathbf{t < 44 \div 8}$ — Divide both sides by 8.

$\mathbf{t < 5.5}$

(c) $k - 17 \le -6$

$\mathbf{k \le -6 + 17}$ — Add 17 to both sides.

$\mathbf{k \le 11}$

(d) $\frac{p}{4} \ge 2$

$\mathbf{p \ge 2 \times 4}$ — Multiply both sides by 4.

$\mathbf{p \ge 8}$

(e) $6s \ne 108$

$\mathbf{s \ne 108 \div 6}$

$\mathbf{s \ne 18}$

(f) $16 + t < 7$

$\mathbf{t < 7 - 16}$

$\mathbf{t < -9}$

(g) $\frac{a}{5} > -9$

$\mathbf{a > -9 \times 5}$

$\mathbf{a > -45}$

(h) $u - 5 < -19$

$\mathbf{u < -19 + 5}$

$\mathbf{u < -14}$

(i) $3m + 5 \le 32$

$\mathbf{3m \le 32 - 5}$ — First subtract 5 from both sides of the inequation,

$\mathbf{3m \le 27}$

$\mathbf{m \le 27 \div 3}$ — then divide both sides by 3.

$\mathbf{m \le 9}$

(j) $2k - 7 \ge 18$

$\mathbf{2k \ge 18 + 7}$ — First add 7 to both sides,

$\mathbf{2k \ge 25}$

$\mathbf{k \ge 25 \div 2}$ — then divide both sides by 2.

$\mathbf{k \ge 12.5}$

(k) $5b - 3 \ne -7$

$\mathbf{5b \ne -7 + 3}$

$\mathbf{5b \ne -4}$

$\mathbf{b \ne -4 \div 5}$

$\mathbf{b \ne -0.8}$ — $-\frac{4}{5}$ is the same thing as -0.8.

(l) $9y + 4 \le -5$

$\mathbf{9y \le -5 - 4}$

$\mathbf{9y \le -9}$

$\mathbf{y \le -9 \div 9}$

$\mathbf{y \le -1}$

2. Graph on the number line provided:

(a) $x \geq -4$

☞ Circle the position at –4.
This circle is solid because –4 is included.
The line points to values larger than –4.

(b) $x < 1$

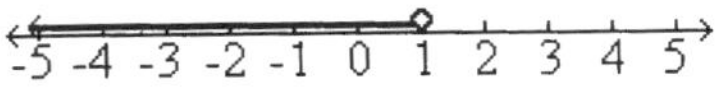

☞ Circle the position at 1.
The circle is open because 1 is not included.
The line points to values smaller than 1.

(c) $x > -1$

☞ Circle the position at –1.
The circle is open because –1 is not included.
The line points to values larger than –1.

(d) $x \leq 3$

☞ Circle the position at 3.
The circle is solid because 3 is included.
The line points to values smaller than 3.

(e) $x < -2$

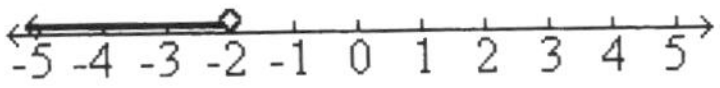

(f) $x \geq 0$

3. Solve and graph the solution on the number line provided:

(a) $2x + 1 \leq 9$

$\mathbf{2x \leq 9 - 1}$

$\mathbf{2x \leq 8}$

$\mathbf{x \leq 8 \div 2}$

$\mathbf{x \leq 4}$

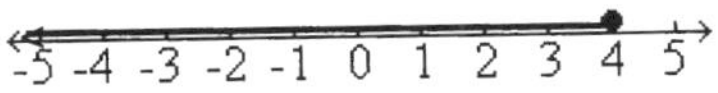

(b) $3x - 7 > -1$

$\mathbf{3x > -1 + 7}$

$\mathbf{3x > 6}$

$\mathbf{x > 6 \div 3}$

$\mathbf{x > 2}$

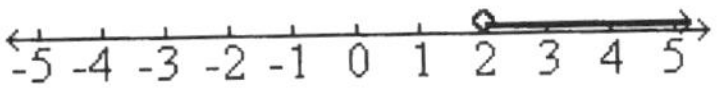

(c) $4x - 4 > -4$

$\mathbf{4x > -4 + 4}$

$\mathbf{4x > 0}$

$\mathbf{x > 0 \div 4}$

$\mathbf{x > 0}$

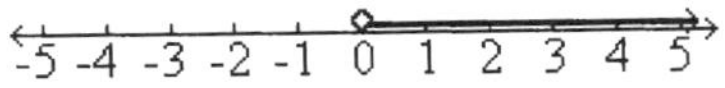

(d) $6x \geq 4x + 8$

$\mathbf{6x - 4x \geq 8}$

$\mathbf{2x \geq 8}$

$\mathbf{x \geq 8 \div 2}$

$\mathbf{x \geq 4}$

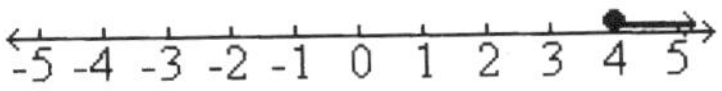

We know that 4 is greater than 3, and we also know that –4 is less than –3.

$4 \times 1 > 3 \times 1$ but $4 \times -1 < 3 \times -1$
$4 \times 2 > 3 \times 2$ but $4 \times -2 < 3 \times -2$
$4 \times 10 > 3 \times 10$ but $4 \times -10 < 3 \times -10$

> When multiplying or dividing both sides of an inequality by a negative number, the inequality sign must be reversed.

Example:

Solve:

(a) $-x \geq 8$ **(b)** $-3m < 12$ **(c)** $\frac{-d}{4} > -8$

Solution:

(a) $-x \geq 8$
$x \leq 8 \div -1$
$x \leq -8$

(b) $-3m < 12$
$m > 12 \div -3$
$m > -4$

(c) $\frac{-d}{4} > -8$
$d < -8 \times -4$
$d < 32$

It is only when we multiply or divide by a negative number that the inequality sign is reversed. If we multiply or divide by a positive number, or add or subtract any number, the inequality sign remains the same.

Example:

Solve:

(a) $5 - 3x \leq 11$ **(b)** $9a - 4 > 14$ **(c)** $2x + 1 < 7x + 11$

Solution:

(a) $5 - 3x \leq 11$
$-3x \leq 6$
$x \geq -2$

(b) $9a - 4 > 14$
$9a > 18$
$a > 2$

(c) $2x + 1 < 7x + 11$
$2x < 7x + 10$
$-5x < 10$
$x > -2$

Example:

Solve and graph the solution on a number line:

(a) $4 - 5x > 19$ **(b)** $8x + 3 \geq 25 - 3x$

Solution:

(a) $4 - 5x > 19$
$-5x > 15$
$x < -3$

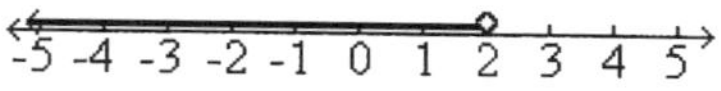

(b) $8x + 3 \geq 25 - 3x$
$8x \geq 22 - 3x$
$11x \geq 22$
$x \geq 2$

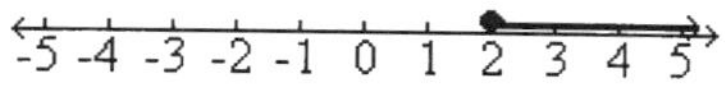

1. Answer true or false:

(a) $7 > 6$
(d) $-7 > 6$
(g) $3 < 2$
(j) $-2 < -3$
(m) $-5 > 4$

(b) $7 > -6$
(e) $-2 < 3$
(h) $-3 < 2$
(k) $5 < -4$
(n) $5 < -4$

(c) $-7 > -6$
(f) $2 < 3$
(i) $-3 < -2$
(l) $-4 > -5$
(o) $4 > -5$

2. Solve:

(a) $-3d \geq 9$
(c) $-e > -3$
(e) $-x \geq 5$
(g) $\frac{-n}{4} \leq 7$

(b) $-8z \leq -8$
(d) $-t < 4$
(f) $\frac{m}{-3} < 15$
(h) $-4y \geq -16$

3. Solve:

(a) $-\frac{b}{2} > -8$
(c) $5t < -20$
(e) $\frac{a}{8} \geq -12$
(g) $3r < 24$

(b) $-h < -6$
(d) $\frac{s}{-3} \leq 12$
(f) $-6d > -24$
(h) $\frac{k}{-5} > 0$

4. Solve:

(a) $8 - 2x > 36$
(b) $3p \leq 24 + 9p$
(c) $9t - 2 < -11$
(d) $-6y + 4 \geq 22$

5. Solve and graph the solution on the number line provided:

(a) $3 - 7x \geq -25$

-5 -4 -3 -2 -1 0 1 2 3 4 5

(b) $6x < 11x + 15$

-5 -4 -3 -2 -1 0 1 2 3 4 5

1. Answer true or false:

(a) $7 > 6$ **True**

(b) $7 > -6$ **True**

(c) $-7 > -6$ **False**

(d) $-7 > 6$ **False**

(e) $-2 < 3$ **True**

(f) $2 < 3$ **True**

(g) $3 < 2$ **False**

(h) $-3 < 2$ **True**

(i) $-3 < -2$ **True**

(j) $-2 < -3$ **False**

(k) $5 < -4$ **False**

(l) $-4 > -5$ **True**

(m) $-5 > 4$ **False**

(n) $5 < -4$ **False**

(o) $4 > -5$ **True**

2. Solve:

(a) $-3d \geq 9$

$\mathbf{d \leq 9 \div -3}$

$\mathbf{d \leq -3}$

☞ We are dividing by –3, so the inequality sign is reversed.

(b) $-8z \leq -8$

$\mathbf{z \geq -8 \div -8}$

$\mathbf{z \geq 1}$

☞ We are dividing by –8; the inequality sign must be reversed.

(c) $-e > -3$

$\mathbf{e < -3 \div -1}$

$\mathbf{e < 3}$

☞ –*e* means –1*e*
Dividing by –1, the inequality sign reverses.

(d) $-t < 4$

$\mathbf{t > 4 \div -1}$

$\mathbf{t > -4}$

☞ Dividing both sides of the inequation by –1.

(e) $-x \geq 5$

$\mathbf{x \leq 5 \div -1}$

$\mathbf{x \leq -5}$

☞ The inequality sign must be reversed.

(f) $\frac{m}{-3} < 15$

$\mathbf{m > 15 \times -3}$

$\mathbf{m > -45}$

☞ We are multiplying by –3 so the inequality sign is reversed.

(g) $\frac{-n}{4} \leq 7$

$\mathbf{n \geq 7 \times -4}$

$\mathbf{n \geq -28}$

☞ Multiplying by –4, the inequality sign must be reversed.

(h) $-4y \geq -16$

$\mathbf{y \leq -16 \div -4}$

$\mathbf{y \leq 4}$

☞ Dividing by –4.

3. Solve:

(a) $-\frac{b}{2} > -8$

$\mathbf{b < -8 \times -2}$ ☞ Multiplying both sides by –2.

$\mathbf{b < 16}$

(b) $-h < -6$

$\mathbf{h > -6 \div -1}$ ☞ Dividing both sides by –1.

$\mathbf{h > 6}$

(c) $5t < -20$

$\mathbf{t < -20 \div 5}$ ☞ We are dividing by a positive number.

$\mathbf{t < -4}$ The inequality sign does not change.

(d) $\frac{s}{-3} \leq 12$

$\mathbf{s \geq 12 \times -3}$ ☞ Multiplying both sides by –3.

$\mathbf{s \geq -36}$

(e) $\frac{a}{8} \geq -12$

$\mathbf{a \geq -12 \times 8}$

$\mathbf{a \geq -96}$

(f) $-6d > -24$

$\mathbf{d < -24 \div -6}$

$\mathbf{d < 4}$

(g) $3r < 24$

$\mathbf{r < 24 \div 3}$

$\mathbf{r < 8}$

(h) $\frac{-k}{5} > 0$

$\mathbf{k < 0 \times -5}$

$\mathbf{k < 0}$

4. Solve:

(a) $8 - 2x > 36$

$\mathbf{-2x > 36 - 8}$ ☞ First subtract 8 from both sides of the inequation,

$\mathbf{-2x > 28}$

$\mathbf{x < 28 \div -2}$ then divide by –2, remembering to reverse the sign.

$\mathbf{x < -14}$

(b) $3p \leq 24 + 9p$

$\mathbf{3p - 9p \leq 24}$

$\mathbf{-6p \leq 24}$

$\mathbf{p \geq 24 \div -6}$

$\mathbf{p \geq -4}$

(c) $9t - 2 < -11$

$\mathbf{9t < -11 + 2}$

$\mathbf{9t < -9}$

$\mathbf{t < -9 \div 9}$

$\mathbf{t < -1}$

(d) $-6y + 4 \geq 22$

$\mathbf{-6y \geq 22 - 4}$

$\mathbf{-6y \geq 18}$

$\mathbf{y \leq 18 \div -6}$

$\mathbf{y \leq -3}$

5. Solve and graph the solution on the number line provided:

(a) $3 - 7x \geq -25$

$\mathbf{-7x \geq -25 - 3}$

$\mathbf{-7x \geq -28}$

$\mathbf{x \leq -28 \div -7}$

$\mathbf{x \leq 4}$

(b) $6x < 11x + 15$

$\mathbf{6x - 11x < 15}$

$\mathbf{-5x < 15}$

$\mathbf{x > 15 \div -5}$

$\mathbf{x > -3}$

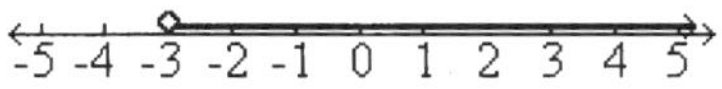

FORMULAS

A formula is a rule that uses pronumerals.
These pronumerals often represent some sort of measurement (such as length, height, mass or period of time).
We use formulas in maths, (finding areas and perimeters, for example), and in science and some other subjects as well.

Many questions that use formulas simply involve substituting.

Example:

Given that $D = MV$, find the value of D when $M = 3.7$ and $V = 5$.

Solution:

$D = MV$
$= 3.7 \times 5$
$= 18.5$

Example:

The formula $C = 2\pi r$ is used to find the circumference of a circle.
Find C, correct to 2 decimal places, when $r = 8$.

Solution:

$C = 2\pi r$
$= 2 \times \pi \times 8$
$= 50.265\ 482\ 46$ [by calculator]
$= 50.27$ [correct to 2 decimal places]

We don't need to understand the purpose of the formula to do the substitution.

Example:

Find the value of S if $S = \frac{n}{2}[2a + (n - 1)d]$, and $n = 14$, $a = 9$ and $d = 7$.

Solution:

$S = \frac{n}{2}[2a + (n - 1)d]$
$= \frac{14}{2}[2 \times 9 + (14 - 1) \times 7]$
$= 7[18 + 13 \times 7]$
$= 7[18 + 91]$
$= 7 \times 109$
$= 763$

We know that $2^2 = 4$ and $(-2)^2 = 4$, so if $x^2 = 4$, $x = 2$ and $x = -2$ are both solutions.
We solve equations like $x^2 = 4$ by taking the square root.

$x^2 = 4$
$x = \sqrt{4}$ or $x = -\sqrt{4}$
$x = 2$ or $x = -2$

In many formulas, pronumerals represent physical measurements that cannot be negative.

Example:

Given the formula $v^2 = u^2 + 2as$, find v when $u = 10$, $a = 7.5$ and $s = 57.4$. ($v > 0$)

Solution:

$v^2 = u^2 + 2as$
$= 10^2 + 2 \times 7.5 \times 57.4$
$= 100 + 861$
$= 961$
$v = \sqrt{961}$ ($v > 0$)
$= 31$

If we had not been told that $v > 0$, there would be two possible solutions.
$v = 31$ or $v = -31$

1. Given that $D = ST$, find the value of D, when:
(a) $S = 80$ and $T = 9$ (b) $S = 56$ and $T = 3\frac{1}{2}$ (c) $S = 67$ and $T = 5$

2. Given that $A = \frac{1}{2}bh$, find A when:
(a) $b = 12$ and $h = 6$ (b) $b = 42$ and $h = 18$ (c) $b = 15$ and $h = 17$

3. Given that $T = a + (n - 1)d$, find T when:
(a) $a = 6$, $n = 8$ and $d = 3$ (b) $a = 11$, $n = 12$ and $d = 20$ (c) $a = 1$, $n = 15$ and $d = 4$

4. Given that $C = \frac{5}{9}(F - 32)$, find C when:
(a) $F = 212$ (b) $F = 32$ (c) $F = 104$

5. Find the value of V, correct to two decimal places, if $V = \sqrt{2gr}$ and $g = 9.8$ and $r = 6400$.

6. The formula $I = \frac{Prn}{100}$ is used to calculate the amount of simple interest. Find I when:
a) $P = 1000$, $r = 6$ and $n = 2$ (b) $P = 7000$, $r = 8$ and $n = 5$ (c) $P = 12\,500$, $r = 9$ and $n = 7$

7. Given the formula $e^2 = 1 - \frac{b^2}{a^2}$, $(e > 0)$, find e when $a = 5$ and $b = 3$.

1. Given that $D = ST$, find the value of D, when:
   (a) $S = 80$ and $T = 9$

$\boldsymbol{D = ST}$ ☞ First write the formula,
$\boldsymbol{D = 80 \times 9}$ then substitute the given values,
$\boldsymbol{D = 720}$ and write down the answer.

   (b) $S = 56$ and $T = 3\frac{1}{2}$

$\boldsymbol{D = ST}$ ☞ Remember, *ST* means $S \times T$.
$\boldsymbol{D = 56 \times 3\frac{1}{2}}$
$\boldsymbol{D = 196}$

   (c) $S = 67$ and $T = 5$

$\boldsymbol{D = ST}$
$\boldsymbol{D = 67 \times 5}$
$\boldsymbol{D = 335}$

2. Given that $A = \frac{1}{2}bh$, find A when:
   (a) $b = 12$ and $h = 6$

$\boldsymbol{A = \frac{1}{2}bh}$ ☞ Write the formula,
$\boldsymbol{A = \frac{1}{2} \times 12 \times 6}$ substitute, and
$\boldsymbol{A = 36}$ write down the answer.

   (b) $b = 42$ and $h = 18$

$\boldsymbol{A = \frac{1}{2}bh}$
$\boldsymbol{A = \frac{1}{2} \times 42 \times 18}$
$\boldsymbol{A = 378}$

   (c) $b = 15$ and $h = 17$

$\boldsymbol{A = \frac{1}{2}bh}$
$\boldsymbol{A = \frac{1}{2} \times 15 \times 17}$
$\boldsymbol{A = 127\frac{1}{2}}$

3. Given that $T = a + (n - 1)d$, find T when:
   (a) $a = 6$, $n = 8$ and $d = 3$

$\boldsymbol{T = a + (n - 1)d}$ ☞ Write the formula,
$\boldsymbol{T = 6 + (8 - 1) \times 3}$ then substitute.
$\boldsymbol{T = 6 + 7 \times 3}$
$\boldsymbol{T = 6 + 21}$
$\boldsymbol{T = 27}$

You might not need to write all these steps.
You could just write down the answer after substituting.

   (b) $a = 11$, $n = 12$ and $d = 20$

$\boldsymbol{T = a + (n - 1)d}$
$\boldsymbol{T = 11 + (12 - 1) \times 20}$
$\boldsymbol{T = 11 + 11 \times 20}$
$\boldsymbol{T = 11 + 220}$
$\boldsymbol{T = 231}$

   (c) $a = 1$, $n = 15$ and $d = 4$

$\boldsymbol{T = a + (n - 1)d}$
$\boldsymbol{T = 1 + (15 - 1) \times 4}$
$\boldsymbol{T = 1 + 14 \times 4}$
$\boldsymbol{T = 1 + 56}$
$\boldsymbol{T = 57}$

4. Given that $C = \frac{5}{9}(F - 32)$, find C when:

(a) $F = 212$

$\mathbf{C = \frac{5}{9}(F - 32)}$

$\mathbf{C = \frac{5}{9}(212 - 32)}$

$\mathbf{C = \frac{5}{9} \times 180}$

$\mathbf{C = 100}$

(b) $F = 32$

$\mathbf{C = \frac{5}{9}(F - 32)}$

$\mathbf{C = \frac{5}{9}(32 - 32)}$

$\mathbf{C = \frac{5}{9} \times 0}$

$\mathbf{C = 0}$

(c) $F = 104$

$\mathbf{C = \frac{5}{9}(F - 32)}$

$\mathbf{C = \frac{5}{9}(104 - 32)}$

$\mathbf{C = \frac{5}{9} \times 72}$

$\mathbf{C = 40}$

5. Find the value of V, correct to two decimal places, if $V = \sqrt{2gr}$ and $g = 9.8$ and $r = 6400$.

$\mathbf{V = \sqrt{2gr}}$

When $g = 9.8$ and $r = 6400$, $\mathbf{V = \sqrt{2 \times 9.8 \times 6400}}$

$\mathbf{V = \sqrt{125\,440}}$

$\mathbf{V = 354.175\,0979}$ [by calculator]

$\mathbf{V = 354.18}$ [correct to two decimal places]

6. The formula $I = \frac{Prn}{100}$ is used to calculate the amount of simple interest. Find I when:

(a) $P = 1000$, $r = 6$ and $n = 2$

$\mathbf{I = \frac{Prn}{100}}$ ☞ Write the formula.

$\mathbf{I = \frac{1000 \times 6 \times 2}{100}}$ Substitute.

$\mathbf{I = \frac{10\not{0}\not{0} \times 6 \times 2}{1\not{0}\not{0}}}$ Cancel before multiplying to simplify.

$\mathbf{I = 120}$

(b) $P = 7000$, $r = 8$ and $n = 5$

$\mathbf{I = \frac{Prn}{100}}$

$\mathbf{I = \frac{7000 \times 8 \times 5}{100}}$

$\mathbf{I = 2800}$

(c) $P = 12\,500$, $r = 9$ and $n = 7$

$\mathbf{I = \frac{Prn}{100}}$

$\mathbf{I = \frac{12\,500 \times 9 \times 7}{100}}$

$\mathbf{I = 7875}$

7. Given the formula $e^2 = 1 - \frac{b^2}{a^2}$, $(e > 0)$, find e when $a = 5$ and $b = 3$.

$\mathbf{e^2 = 1 - \frac{b^2}{a^2}}$ ☞ Write the formula.

$\mathbf{e^2 = 1 - \frac{3^2}{5^2}}$ Substitute.

$\mathbf{e^2 = 1 - \frac{9}{25}}$

$\mathbf{e^2 = \frac{16}{25}}$

$\mathbf{e = \sqrt{\frac{16}{25}}}$

$\mathbf{e = \frac{4}{5}}$

You could finish this using decimals instead of fractions:

$\mathbf{e^2 = 0.64}$

$\mathbf{e = \sqrt{0.64}}$

$\mathbf{e = 0.8}$

EQUATIONS ARISING FROM FORMULAS

The pronumeral found before the equals sign is called the **subject** of the formula.
When we know the value of the subject of the formula, but want to find one of the other values, we need to solve an equation.

Example:

Given $A = lb$, find:

(a) b if $A = 28$ and $l = 7$

(b) l if $A = 76$ and $b = 4$

Solution:

(a)
$$A = lb$$
$$28 = 7 \times b$$
$$7b = 28$$
$$b = 28 \div 7$$
$$b = 4$$

(b)
$$A = lb$$
$$76 = l \times 4$$
$$4l = 76$$
$$l = 76 \div 4$$
$$l = 19$$

We can always turn an equation around so that the pronumeral we are trying to find is at the front.

Example:

Given $v = u + at$, find:

(a) u if $v = 56$, $a = 10$ and $t = 5$

(b) a if $v = 49$, $u = 5$ and $t = 4$

Solution:

(a)
$$v = u + at$$
$$56 = u + 10 \times 5$$
$$u + 50 = 56$$
$$u = 56 - 50$$
$$u = 6$$

(b)
$$v = u + at$$
$$49 = 5 + a \times 4$$
$$5 + 4a = 49$$
$$4a = 44$$
$$a = 11$$

Many formulas involve squares. Remember that we solve these by taking the square root.

Example:

(a) Given $V = x^2y$, find x if $V = 605$ and $y = 5$, $(x > 0)$

(b) If $c^2 = a^2 + b^2$, and $c = 37$ and $b = 35$, find a, $(a > 0)$

Solution:

(a)
$$V = x^2y$$
$$605 = x^2 \times 5$$
$$5x^2 = 605$$
$$x^2 = 605 \div 5$$
$$x^2 = 121$$
$$x = \sqrt{121} \quad (x > 0)$$
$$x = 11$$

(b)
$$c^2 = a^2 + b^2$$
$$37^2 = a^2 + 35^2$$
$$1369 = a^2 + 1225$$
$$a^2 = 1369 - 1225$$
$$a^2 = 144$$
$$a = \sqrt{144} \quad (a > 0)$$
$$a = 12$$

1. Given that $V = Ah$, find the value of:
   (a) A if $V = 96$ and $h = 4$
   (b) h if $V = 855$ and $A = 95$

2. Given that $S = \frac{D}{T}$, find:
   (a) D when $S = 60$ and $T = 5$
   (b) T when $S = 72$ and $D = 252$

3. Given that $v^2 = u^2 + 2as$, find:
   (a) s when $v = 15$, $u = 11$ and $a = 4$
   (b) u when $v = 20$, $a = 8$ and $s = 16$ $(u > 0)$

4. Given that $A = \frac{1}{2}xy$, find:
   (a) y when $A = 54$ and $x = 12$
   (b) x when $A = 21$ and $y = 7$

5. Given that $S = 6x^2$, $(x > 0)$, find x when:
   (a) $S = 96$
   (b) $S = 216$

EQUATIONS ARISING FROM FORMULAS Worked Solutions

1. Given that $V = Ah$, find the value of:

(a) A if $V = 96$ and $h = 4$

$\mathbf{V = Ah}$	Write the formula.
$\mathbf{96 = A \times 4}$	Substitute the given values.
$\mathbf{4A = 96}$	Turn the equation around.
$\mathbf{A = 96 \div 4}$	Divide both sides of the equation by 4.
$\mathbf{A = 24}$	

(b) h if $V = 855$ and $A = 95$

$\mathbf{V = Ah}$	Always write the formula first.
$\mathbf{855 = 95 \times h}$	
$\mathbf{95h = 855}$	
$\mathbf{h = 855 \div 95}$	
$\mathbf{h = 9}$	

2. Given that $S = \frac{D}{T}$, find:

(a) D when $S = 60$ and $T = 5$

$\mathbf{S = \frac{D}{T}}$	Write the formula.
$\mathbf{60 = \frac{D}{5}}$	Substitute the given values.
$\mathbf{\frac{D}{5} = 60}$	
$\mathbf{D = 60 \times 5}$	Multiply both sides of the equation by 5.
$\mathbf{D = 300}$	

(b) T when $S = 72$ and $D = 252$

$\mathbf{S = \frac{D}{T}}$	
$\mathbf{72 = \frac{252}{T}}$	In this case we don't need to turn the equation around.
$\mathbf{72T = 252}$	We multiply both sides by T instead.
$\mathbf{T = 252 \div 72}$	Divide both sides by 72.
$\mathbf{T = 3.5}$	

3. Given that $v^2 = u^2 + 2as$, find:

(a) s when $v = 15$, $u = 11$ and $a = 4$

$\mathbf{v^2 = u^2 + 2as}$	Write the formula.
$\mathbf{15^2 = 11^2 + 2 \times 4 \times s}$	Substitute.
$\mathbf{225 = 121 + 8s}$	
$\mathbf{8s + 121 = 225}$	
$\mathbf{8s = 225 - 121}$	Subtracting 121 from both sides of the equation.
$\mathbf{8s = 104}$	
$\mathbf{s = 104 \div 8}$	Dividing both sides by 8.
$\mathbf{s = 13}$	

(b) u when $v = 20$, $a = 8$ and $s = 16$ $(u > 0)$

$$v^2 = u^2 + 2as$$

☞ Write the formula.

$$20^2 = u^2 + 2 \times 8 \times 16$$

Substitute.

$$400 = u^2 + 256$$

$$u^2 + 256 = 400$$

Turn the equation around.

$$u^2 = 400 - 256$$

$$u^2 = 144$$

$$u = \sqrt{144} \quad (u > 0)$$

$$u = 12$$

Since we are told that $u > 0$, there is only one solution.

4. Given that $A = \frac{1}{2}xy$, find:

(a) y when $A = 54$ and $x = 12$

$$A = \frac{1}{2}xy$$

$$54 = \frac{1}{2} \times 12 \times y$$

$$54 = 6y$$

☞ On multiplying. ($\frac{1}{2}$ of 12 = 6)

$$6y = 54$$

$$y = 54 \div 6$$

$$y = 9$$

(b) x when $A = 21$ and $y = 7$

$$A = \frac{1}{2}xy$$

$$21 = \frac{1}{2} \times x \times 7$$

$$\frac{7}{2}x = 21$$

$$\overset{1}{\cancel{2}} \times \frac{7}{\underset{1}{\cancel{2}}}x = 2 \times 21$$

☞ Multiplying both sides by the common denominator.

$$7x = 42 \qquad \text{or} \qquad \frac{7}{2}x = 21$$

$$x = 42 \div 7 \qquad\qquad x = 21 \div \frac{7}{2}$$

$$x = 6 \qquad\qquad x = 6$$

5. Given that $S = 6x^2$, $(x > 0)$, find x when:

(a) $S = 96$

$$S = 6x^2$$

$$96 = 6x^2$$

$$6x^2 = 96$$

$$x^2 = 96 \div 6$$

$$x^2 = 16$$

$$x = \sqrt{16} \quad (x > 0)$$

$$x = 4$$

(b) $S = 216$

$$S = 6x^2$$

$$216 = 6x^2$$

$$6x^2 = 216$$

$$x^2 = 216 \div 6$$

$$x^2 = 36$$

$$x = \sqrt{36} \quad (x > 0)$$

$$x = 6$$

82 CHANGING THE SUBJECT OF A FORMULA

Remember: The subject of a formula is the part before the '=' sign.
When changing the subject of a formula, we simply rearrange it, using the rules that we use to solve equations.

> It often helps to first turn the formula around!

Example:

Make D the subject of the formula $S = \frac{D}{T}$.

Solution:

$S = \frac{D}{T}$

$\frac{D}{T} = S$

$D = ST$ (Multiplying both sides by T.)

Example:

Given the formula, $P = 2L + 2B$, change the subject to B.

Solution:

$P = 2L + 2B$

$2L + 2B = P$ (Writing the formula the other way around.)

$2B = P - 2L$ (Subtracting $2L$ from both sides.)

$B = \frac{P - 2L}{2}$ (Dividing both sides by 2.)

> There are often different ways of expressing the answer. For example, we could have put $B = \frac{P}{2} - L$ in this solution.

Example:

The formula for finding the area of a trapezium is $A = \frac{h}{2}(a + b)$. Make:

(a) h the subject of this formula **(b) b the subject of this formula.**

Solution:

(a)

$A = \frac{h}{2}(a + b)$

$\frac{h}{2}(a + b) = A$

$h(a + b) = 2A$

$h = \frac{2A}{a + b}$

(b)

$A = \frac{h}{2}(a + b)$

$\frac{h}{2}(a + b) = A$

$h(a + b) = 2A$

$a + b = \frac{2A}{h}$

$b = \frac{2A}{h} - a$

Example:

Given the formula, $S = \frac{a}{1 - r}$, change the subject to r.

> There can also be different ways to do the same question.

Solution:

$S = \frac{a}{1 - r}$

$S(1 - r) = a$ (Multiplying by $(1 - r)$)

$1 - r = \frac{a}{S}$ (Dividing by S)

$-r = \frac{a}{S} - 1$ (Subtracting 1)

$r = 1 - \frac{a}{S}$ (Dividing by -1)

or

$S = \frac{a}{1 - r}$

$\frac{a}{1 - r} = S$ (Turning the formula around)

$a = S - Sr$ (Multiplying by $(1 - r)$)

$a + Sr = S$ (Adding Sr)

$Sr = S - a$ (Subtracting a)

$r = 1 - \frac{a}{S}$ (Dividing by S)

1. Change the subject of the formula $V = IR$ to R.

2. Given the formula $T = a + (n - 1)d$, change the subject to:
   (a) a
   (b) n

3. Given the formula $A = \frac{b + c}{2}$, change the subject to c.

4. The formula for the volume of a cone is $V = \frac{1}{3}\pi r^2 h$. Change the subject of this formula to:
   (a) h
   (b) r $(r > 0)$

5. The cost to hire a car is given by the formula $C = 20 + \frac{3}{4}d$. Make d the subject of this formula.

6. Given the formula $S = \frac{n}{2}(a + l)$, change the subject to:
   (a) n
   (b) l

84 CHANGING THE SUBJECT OF A FORMULA Worked Solutions

1. Change the subject of the formula $V = IR$ to R.

$\boldsymbol{V = IR}$ — Write the formula.

$\boldsymbol{IR = V}$ — Turning the equation around.

$\boldsymbol{R = \frac{V}{I}}$ — Dividing both sides by I.

2. Given the formula $T = a + (n-1)d$, change the subject to:

(a) a

$\boldsymbol{T = a + (n-1)d}$ — Write the formula.

$\boldsymbol{a + (n-1)d = T}$ — Turn it around.

$\boldsymbol{a = T - (n-1)d}$ — Subtract $(n-1)d$ from both sides of the equation. $\boldsymbol{a = T - nd + d}$ is the same thing, but it isn't necessary to expand the bracket.

(b) n

$\boldsymbol{T = a + (n-1)d}$

$\boldsymbol{a + (n-1)d = T}$

$\boldsymbol{(n-1)d = T - a}$ — Subtracting a from both sides of the equation.

$\boldsymbol{n - 1 = \frac{T-a}{d}}$ — Dividing both sides by d.

$\boldsymbol{n = \frac{T-a}{d} + 1}$ — Adding 1 to both sides of the equation. $n = \frac{T-a+d}{d}$ is the same thing.

3. Given the formula $A = \frac{b+c}{2}$, change the subject to c.

$\boldsymbol{A = \frac{b+c}{2}}$

$\boldsymbol{\frac{b+c}{2} = A}$ — Turning the formula around.

$\boldsymbol{{}^{1}\cancel{2} \times \left(\frac{b+c}{\cancel{2}_{1}}\right) = 2 \times A}$ — Multiplying both sides of the equation by 2.

$\boldsymbol{b + c = 2A}$

$\boldsymbol{c = 2A - b}$ — Subtracting b from both sides.

4. The formula for the volume of a cone is $V = \frac{1}{3}\pi r^2 h$. Change the subject of this formula to:

(a) h

$\boldsymbol{V = \frac{1}{3}\pi r^2 h}$

$\boldsymbol{\frac{1}{3}\pi r^2 h = V}$ — Turning the formula around.

$\boldsymbol{{}^{1}\cancel{3} \times \frac{1}{\cancel{3}_{1}}\pi r^2 h = 3 \times V}$ — Multiplying both sides by 3.

$\boldsymbol{\pi r^2 h = 3V}$

$\boldsymbol{h = \frac{3V}{\pi r^2}}$ — Dividing both sides by πr^2.

(b) r $(r > 0)$

$$V = \frac{1}{3}\pi r^2 h$$

$$\frac{1}{3}\pi r^2 h = V$$

$$^{1}\cancel{3} \times \frac{1}{\cancel{3}_{1}}\pi r^2 h = 3 \times V$$

Multiplying both sides by 3.

$$\pi r^2 h = 3V$$

$$r^2 = \frac{3V}{\pi h}$$

Dividing both sides by πh.

$$r = \sqrt{\frac{3V}{\pi h}} \quad (r > 0)$$

Taking the positive square root.

5. The cost to hire a car is given by the formula $C = 20 + \frac{3}{4}d$. Make d the subject of this formula.

$$C = 20 + \frac{3}{4}d$$

Write the formula.

$$20 + \frac{3}{4}d = C$$

$$\frac{3}{4}d = C - 20$$

Subtracting 20 from both sides.

$$3d = 4(C - 20)$$

Multiplying both sides by 4.

$$d = \frac{4}{3}(C - 20)$$

Dividing both sides by 3.

You could have combined these last two steps and just divided by $\frac{3}{4}$, (or multiplied by $\frac{4}{3}$).
$d = \frac{4C - 80}{3}$ is the same thing.

6. Given the formula $S = \frac{n}{2}(a + l)$, change the subject to:

(a) n

$$S = \frac{n}{2}(a + l)$$

$$\frac{n}{2}(a + l) = S$$

$$^{1}\cancel{2} \times \frac{n}{\cancel{2}_{1}}(a + l) = 2 \times S$$

Multiplying both sides by 2.

$$n(a + l) = 2S$$

$$n = \frac{2S}{a + l}$$

Dividing both sides by $(a + l)$.

(b) l

$$S = \frac{n}{2}(a + l)$$

$$\frac{n}{2}(a + l) = S$$

$$^{1}\cancel{2} \times \frac{n}{\cancel{2}_{1}}(a + l) = 2 \times S$$

$$n(a + l) = 2S$$

$$a + l = \frac{2S}{n}$$

Dividing both sides by n.

$$l = \frac{2S}{n} - a$$

Subtracting a from both sides.

common factor is a number, or term, that divides into every term of an expression. When an algebraic expression has a common factor, we can **factorise** by writing the expression as a product of its factors.

Example:

Factorise:

(a) $2x + 14$
(b) $3p - 15q$
(c) $15 - 20z$
(d) $12x + 18y + 6$
(e) $x^2 - 6x$
(f) $ab + 9b$
(g) $xy - x$
(h) $4t - tv + tw$

Solution:

(a) $2x + 14 = \mathbf{2(x + 7)}$
(b) $3p - 15q = \mathbf{3(p - 5q)}$
(c) $15 - 20z = \mathbf{5(3 - 4z)}$
(d) $12x + 18y + 6 = \mathbf{6(2x + 3y + 1)}$
(e) $x^2 - 6x = \mathbf{x(x - 6)}$
(f) $ab + 9b = \mathbf{b(a + 9)}$
(g) $xy - x = \mathbf{x(y - 1)}$
(h) $4t - tv + tw = \mathbf{t(4 - v + w)}$

A common factor might be a combination of a number and a term, or of several terms.

Example:

Factorise:

(a) $2x^2 + 8x$
(b) $10pq - 5q$
(c) $12a + 9ab$
(d) $8ab - abc$
(e) $xyz - xy$
(f) $cde + def + 7de$

Solution:

(a) $2x^2 + 8x = \mathbf{2x(x + 4)}$
(b) $10pq - 5q = \mathbf{5q(2p - 1)}$
(c) $12a + 9ab = \mathbf{3a(4 + 3b)}$
(d) $8ab - abc = \mathbf{ab(8 - c)}$
(e) $xyz - xy = \mathbf{xy(z - 1)}$
(f) $cde + def + 7de = \mathbf{de(c + f + 7)}$

Consider $p^5 + 3p^3$.
p is a common factor because it divides into both p^5 and $3p^3$.
p^2 is also a common factor.
But, the largest common factor is p^3.
We must take care to always find the **largest** common factor!

Example:

Factorise:

(a) $p^5 + 3p^3$
(b) $x^3 - x^2$
(c) $9a^2bc^3 + 3abc^2$
(d) $12p^4q^7r^6 - 18p^3q^8r^3$

Solution:

(a) $p^5 + 3p^3 = \mathbf{p^3(p^2 + 3)}$
(b) $x^3 - x^2 = \mathbf{x^2(x - 1)}$
(c) $9a^2bc^3 + 3abc^2 = \mathbf{3abc^2(3ac + 1)}$
(d) $12p^4q^7r^6 - 18p^3q^8r^3 = \mathbf{6p^3q^7r^3(2pr^3 - 3q)}$

1. **Factorise, by taking out a common factor:**

(a) $3x + 27$ = ______
(b) $7a + 35b$ = ______
(c) $20x - 15$ = ______
(d) $45t + 9$ = ______
(e) $6y - 20$ = ______
(f) $8g + 12h$ = ______
(g) $10 - 50b$ = ______
(h) $14c + 49d$ = ______
(i) $2a + 8b - 6c$ = ______
(j) $x^2 + 7x$ = ______
(k) $a^2 - 10a$ = ______
(l) $mn - 6m$ = ______
(m) $pq - q^2$ = ______
(n) $8t + t^2$ = ______
(o) $ab - b$ = ______
(p) $j + jk$ = ______
(q) $24g + 18$ = ______
(r) $28 - 32n$ = ______
(s) $e^2 - 5e + ef$ = ______
(t) $x^2 - xy - xz + 9x$ = ______

2. **Factorise fully:**

(a) $3x^2 - 9x$ = ______
(b) $7a^2 + 49a$ = ______
(c) $6pq + 3p$ = ______
(d) $4x^2 - 2xy$ = ______
(e) $abc + 6bc$ = ______
(f) $15y - 9yz$ = ______
(g) $2h^2 + 2h$ = ______
(h) $m^2n + mn$ = ______
(i) $12d - 3cd$ = ______
(j) $pqr - pq$ = ______
(k) $16t^2 + 12tu$ = ______
(l) $18ab - 24bc$ = ______
(m) $3t - 9t^2$ = ______
(n) $8xy + 2y$ = ______
(o) $10bc - 25b$ = ______
(p) $abc + bcd + cde$ = ______
(q) $4x^2 - 2xy + 8x$ = ______

3. **Factorise:**

(a) $a^3 + a^2$ = ______
(b) $k^5 - k^3$ = ______
(c) $t^4 + t$ = ______
(d) $e^2 - e^6$ = ______
(e) $h^7 + h^4$ = ______
(f) $m^6 - m^3$ = ______
(g) $2b^2 + 4b^3$ = ______
(h) $4ab - 6abc$ = ______
(i) $d^7 - 6d^6$ = ______
(j) $3h^5 + 4h^4$ = ______
(k) $p^5 - pq$ = ______
(l) $z^3 + y^2z$ = ______
(m) $d^8 - cd^2$ = ______
(n) $4n^5 - 6n^3$ = ______
(o) $8k + 12k^3$ = ______
(p) $6a^2b^3 + 9a^3b^7$ = ______
(q) $4x^5y^3 - 8x^3y^2$ = ______
(r) $9p^4q^5 - p^2q^5$ = ______
(s) $c^8d^2 + 7cde$ = ______
(t) $14a^2b^4c - 7abc^3$ = ______
(u) $12pq^3r^3 - 15p^4qr^2$ = ______

COMMON FACTORS — REVIEW Worked Solutions

1. Factorise, by taking out a common factor:

(a) $3x + 27 = 3(x + 9)$ ☞ $3x + 27 = 3 \times x + 3 \times 9$

(b) $7a + 35b = 7(a + 5b)$ ☞ $7a + 35b = 7 \times a + 7 \times 5b$

(c) $20x - 15 = 5(4x - 3)$ ☞ $20x - 15 = 5 \times 4x - 5 \times 3$

(d) $45t + 9 = 9(5t + 1)$ ☞ $45t + 9 = 9 \times 5t + 9 \times 1$

(e) $6y - 20 = 2(3y - 10)$

(f) $8g + 12h = 4(2g + 3h)$ ☞ Always take out the largest common factor.

(g) $10 - 50b = 10(1 - 5b)$ ☞ $10 - 50b = 10 \times 1 - 10 \times 5b$

(h) $14c + 49d = 7(2c + 7d)$ ☞ We can always check the factorisation by expanding.
$7(2c + 7d) = 7 \times 2c + 7 \times 7d$

(i) $2a + 8b - 6c = 2(a + 4b - 3c)$ ☞ $2a + 8b - 6c = 2 \times a + 2 \times 4b - 2 \times 3c$

(j) $x^2 + 7x = x(x + 7)$ ☞ $x^2 + 7x = x \times x + x \times 7$

(k) $a^2 - 10a = a(a - 10)$

(l) $mn - 6m = m(n - 6)$

(m) $pq - q^2 = q(p - q)$ ☞ $pq - q^2 = p \times q - q \times q$

(n) $8t + t^2 = t(8 + t)$

(o) $ab - b = b(a - 1)$ ☞ $ab - b = b \times a - b \times 1$

(p) $j + jk = j(1 + k)$

(q) $24g + 18 = 6(4g + 3)$ ☞ Both 2 and 3 are common factors,
but the largest common factor is 6.

(r) $28 - 32n = 4(7 - 8n)$

(s) $e^2 - 5e + ef = e(e - 5 + f)$ ☞ $e^2 - 5e + ef = e \times e - e \times 5 + e \times f$

(t) $x^2 - xy - xz + 9x = x(x - y - z + 9)$ ☞ Check, by expanding, that the factorisation is correct.

2. Factorise fully:

(a) $3x^2 - 9x = 3x(x - 3)$ ☞ Both 3 and x are common factors.
$3x^2 - 9x = 3x \times x - 3x \times 3$

(b) $7a^2 + 49a = \mathbf{7a(a + 7)}$

(c) $6pq + 3p = \mathbf{3p(2q + 1)}$ ☞ $6pq + 3p = \mathbf{3p \times 2q + 3p \times 1}$

(d) $4x^2 - 2xy = \mathbf{2x(2x - y)}$

(e) $abc + 6bc = \mathbf{bc(a + 6)}$

(f) $15y - 9yz = \mathbf{3y(5 - 3z)}$

(g) $2h^2 + 2h = \mathbf{2h(h + 1)}$

(h) $m^2n + mn = \mathbf{mn(m + 1)}$

(i) $12d - 3cd = \mathbf{3d(4 - c)}$

(j) $pqr - pq = \mathbf{pq(r - 1)}$

(k) $16t^2 + 12tu = \mathbf{4t(4t + 3u)}$

(l) $18ab - 24bc = \mathbf{6b(3a - 4c)}$

(m) $3t - 9t^2 = \mathbf{3t(1 - 3t)}$

(n) $8xy + 2y = \mathbf{2y(4x + 1)}$

(o) $10bc - 25b = \mathbf{5b(2c - 5)}$

(p) $abc + bcd + cde = \mathbf{c(ab + bd + de)}$

(q) $4x^2 - 2xy + 8x = \mathbf{2x(2x - y + 4)}$

3. Factorise:

(a) $a^3 + a^2 = \mathbf{a^2(a + 1)}$ ☞ $a^3 + a^2 = \mathbf{a^2 \times a + a^2 \times 1}$

(b) $k^5 - k^3 = \mathbf{k^3(k^2 - 1)}$ ☞ $k^5 - k^3 = \mathbf{k^3 \times k^2 - k^3 \times 1}$

(c) $t^4 + t = \mathbf{t(t^3 + 1)}$ ☞ $t^4 + t = \mathbf{t \times t^3 + t \times 1}$

(d) $e^2 - e^6 = \mathbf{e^2(1 - e^4)}$

(e) $h^7 + h^4 = \mathbf{h^4(h^3 + 1)}$

(f) $m^6 - m^3 = \mathbf{m^3(m^3 - 1)}$

(g) $2b^2 + 4b^3 = \mathbf{2b^2(1 + 2b)}$ ☞ $2b^2 + 4b^3 = \mathbf{2b^2 \times 1 + 2b^2 \times 2b}$

(h) $4ab - 6abc = \mathbf{2ab(2 - 3c)}$

(i) $d^7 - 6d^6 = \mathbf{d^6(d - 6)}$ ☞ Take out the largest common factor, d^6.

(j) $3h^5 + 4h^4 = \mathbf{h^4(3h + 4)}$

(k) $p^5 - pq = \mathbf{p(p^4 - q)}$

(l) $z^3 + y^2z = \mathbf{z(z^2 + y^2)}$

(m) $d^8 - cd^2 = \mathbf{d^2(d^6 - c)}$

(n) $4n^5 - 6n^3 = \mathbf{2n^3(2n^2 - 3)}$

(o) $8k + 12k^3 = \mathbf{4k(2 + 3k^2)}$

(p) $6a^2b^3 + 9a^3b^7 = \mathbf{3a^2b^3(2 + 3ab^4)}$ ☞ $6a^2b^3 + 9a^3b^7 = \mathbf{3a^2b^3 \times 2 + 3a^2b^3 \times 3ab^4}$

(q) $4x^5y^3 - 8x^3y^2 = \mathbf{4x^3y^2(x^2y - 2)}$ ☞ $4x^5y^3 - 8x^3y^2 = \mathbf{4x^3y^2 \times x^2y - 4x^3y^2 \times 2}$

(r) $9p^4q^5 - p^2q^5 = \mathbf{p^2q^5(9p^2 - 1)}$ ☞ $9p^4q^5 - p^2q^5 = \mathbf{p^2q^5 \times 9p^2 - p^2q^5 \times 1}$

(s) $c^8d^2 + 7cde = \mathbf{cd(c^7d + 7e)}$ ☞ Check the factorisation, by expanding.

(t) $14a^2b^4c - 7abc^3 = \mathbf{7abc(2ab^3 - c^2)}$ ☞ $14a^2b^4c - 7abc^3 = \mathbf{7abc \times 2ab^3 - 7abc \times c^2}$

(u) $12pq^3r^3 - 15p^4qr^2 = \mathbf{3pqr^2(4q^2r - 5p^3)}$ ☞ $12pq^3r^3 - 15p^4qr^2 = \mathbf{3pqr^2 \times 4q^2r - 3pqr^2 \times 5p^3}$

90 FACTORISING BY GROUPING

A common factor could be a sum, difference, or some other combination of two or more terms.

Just as the common factor of $ab + 3b$ is b,
the common factor of $a(p + q) + 3(p + q)$ is $(p + q)$.

$ab + 3b = b(a + 3)$

$a(p + q) + 3(p + q) = (p + q)(a + 3)$

Example:

Factorise:

(a) $x(x + y) + 2(x + y)$ **(b)** $t(u - 3) + 7(u - 3)$ **(c)** $a(b + c) - b(b + c)$

Solution:

(a) $x(x + y) + 2(x + y) = \mathbf{(x + y)(x + 2)}$

(b) $t(u - 3) + 7(u - 3) = \mathbf{(u - 3)(t + 7)}$

(c) $a(b + c) - b(b + c) = \mathbf{(b + c)(a - b)}$

By grouping terms, we can sometimes find a common factor.

$ab + 3b + 4a + 12$
$= b(a + 3) + 4(a + 3)$
$= (a + 3)(b + 4)$

The first 2 terms have a common factor (b) and the last 2 terms have a different common factor (4).

Example:

Factorise, by grouping:

(a) $pq + 6p + qr + 6r$ **(b)** $x^2 - 5x + xy - 5y$ **(c)** $a^2 + 9a + 2ab + 18b$ **(d)** $e^2 - ef + e - f$

Solution:

(a) $pq + 6p + qr + 6r = \mathbf{p(q + 6) + r(q + 6)}$
$= \mathbf{(q + 6)(p + r)}$

(b) $x^2 - 5x + xy - 5y = \mathbf{x(x - 5) + y(x - 5)}$
$= \mathbf{(x - 5)(x + y)}$

(c) $a^2 + 9a + 2ab + 18b = \mathbf{a(a + 9) + 2b(a + 9)}$
$= \mathbf{(a + 9)(a + 2b)}$

(d) $e^2 - ef + e - f = \mathbf{e(e - f) + 1(e - f)}$
$= \mathbf{(e - f)(e + 1)}$

Extra care must be taken with negative factors.

$-2p + 2q = -2(p - q)$

$-x^2 - 7x = -x(x + 7)$

Check, by expanding, that the factorisation is correct.

Example:

Factorise:

(a) $t^2 - tu - 8t + 8u$ **(b)** $ac + ad - bc - bd$

Solution:

(a) $t^2 - tu - 8t + 8u = \mathbf{t(t - u) - 8(t - u)}$
$= \mathbf{(t - u)(t - 8)}$

(b) $ac + ad - bc - bd = \mathbf{a(c + d) - b(c + d)}$
$= \mathbf{(c + d)(a - b)}$

1. Factorise:

(a) $a(a+b)+4(a+b)=$ ______

(b) $a(b+c)+5(b+c)=$ ______

(c) $x(x+y)-7(x+y)=$ ______

(d) $p(p-q)+3(p-q)=$ ______

(e) $6(c-d)-d(c-d)=$ ______

(f) $g(g-4)-4(g-4)=$ ______

(g) $7(t+2)+u(t+2)=$ ______

(h) $k(k-6)-(k-6)=$ ______

2. Factorise by grouping:

(a) $ab+bc+3a+3c$

(b) $gh-3g+ah-3a$

(c) $p^2+pq+pr+qr$

(d) $t^2-tu+8t-8u$

(e) $xy-xz+y^2-yz$

(f) $6c+2d+3bc+bd$

(g) $m^2+4m+mn+4n$

(h) $6a^2+9a+8ab+12b$

(i) $xz+x+yz+y$

(j) $ef+eg+f+g$

(k) $7p+7q+ap^2+apq$

(l) $c^2-cd+c-d$

3. Factorise:

(a) $b^2+bc-3b-3c$

(b) $p^2+pq-pr-qr$

(c) $ef-eg-fh+gh$

(d) $m^2-5m-mn+5n$

(e) $xy+y-xz-z$

(f) $a^2-ab-a+b$

(g) $k^2-kt+2k-2t$

(h) $3x^2+9x+xy+3y$

(i) $ab-ac-bd+cd$

(j) $6t^2+2tu-3t-u$

(k) $abc-abd-c+d$

92 FACTORISING BY GROUPING Worked Solutions

1. Factorise:

(a) $a(a + b) + 4(a + b) = \mathbf{(a + b)(a + 4)}$

$(a + b)$ is the common factor.

(b) $a(b + c) + 5(b + c) = \mathbf{(b + c)(a + 5)}$

$(b + c)$ is the common factor.

(c) $x(x + y) - 7(x + y) = \mathbf{(x + y)(x - 7)}$

(d) $p(p - q) + 3(p - q) = \mathbf{(p - q)(p + 3)}$

$(p - q)$ is the common factor.

(e) $6(c - d) - d(c - d) = \mathbf{(c - d)(6 - d)}$

(f) $g(g - 4) - 4(g - 4) = \mathbf{(g - 4)(g - 4)}$

We can also write the answer as $(g - 4)^2$.

(g) $7(t + 2) + u(t + 2) = \mathbf{(t + 2)(7 + u)}$

(h) $k(k - 6) - (k - 6) = \mathbf{(k - 6)(k - 1)}$

$k(k - 6) - (k - 6) = \mathbf{k(k - 6) - 1(k - 6)}$
We don't need to write the 1 in the question but we MUST write it in the answer.

2. Factorise by grouping:

(a) $ab + bc + 3a + 3c = \mathbf{b(a + c) + 3(a + c)}$

$= \mathbf{(a + c)(b + 3)}$

The first two terms have a common factor (b), the last two terms have a different common factor (3). After taking out the new common factor, $(a + c)$.

(b) $gh - 3g + ah - 3a = \mathbf{g(h - 3) + a(h - 3)}$
$= \mathbf{(h - 3)(g + a)}$

(c) $p^2 + pq + pr + qr = \mathbf{p(p + q) + r(p + q)}$
$= \mathbf{(p + q)(p + r)}$

(d) $t^2 - tu + 8t - 8u = \mathbf{t(t - u) + 8(t - u)}$
$= \mathbf{(t - u)(t + 8)}$

Remember that we can always check, by expanding, that the factorisation is correct.

(e) $xy - xz + y^2 - yz = \mathbf{x(y - z) + y(y - z)}$
$= \mathbf{(y - z)(x + y)}$

(f) $6c + 2d + 3bc + bd = \mathbf{2(3c + d) + b(3c + d)}$
$= \mathbf{(3c + d)(2 + b)}$

(g) $m^2 + 4m + mn + 4n = \mathbf{m(m + 4) + n(m + 4)}$
$= \mathbf{(m + 4)(m + n)}$

(h) $6a^2 + 9a + 8ab + 12b = \mathbf{3a(2a + 3) + 4b(2a + 3)}$
$= \mathbf{(2a + 3)(3a + 4b)}$

Always take out the largest common factor.
($3a$ for 1st 2 terms, $4b$ for last 2 terms.)

(i) $xz + x + yz + y = \mathbf{x(z + 1) + y(z + 1)}$
$= \mathbf{(z + 1)(x + y)}$

☞ $x \times z + x \times 1 + y \times z + y \times 1$
Don't forget the 1.

(j) $ef + eg + f + g = \mathbf{e(f + g) + 1(f + g)}$
$= \mathbf{(f + g)(e + 1)}$

☞ $e(f + g) + (f + g)$ is the same thing.
We can't leave out the 1 in this line.

(k) $7p + 7q + ap^2 + apq = \mathbf{7(p + q) + ap(p + q)}$
$= \mathbf{(p + q)(7 + ap)}$

(l) $c^2 - cd + c - d = \mathbf{c(c - d) + 1(c - d)}$
$= \mathbf{(c - d)(c + 1)}$

3. Factorise:

(a) $b^2 + bc - 3b - 3c = \mathbf{b(b + c) - 3(b + c)}$
$= \mathbf{(b + c)(b - 3)}$

☞ Because we are subtracting the 2nd bracket, the sign inside the bracket will change.

(b) $p^2 + pq - pr - qr = \mathbf{p(p + q) - r(p + q)}$
$= \mathbf{(p + q)(p - r)}$

☞ Be careful with the signs.

(c) $ef - eg - fh + gh = \mathbf{e(f - g) - h(f - g)}$
$= \mathbf{(f - g)(e - h)}$

☞ The sign in the second bracket has changed.

(d) $m^2 - 5m - mn + 5n = \mathbf{m(m - 5) - n(m - 5)}$
$= \mathbf{(m - 5)(m - n)}$

☞ Check, by expanding, that the signs are correct.

(e) $xy + y - xz - z = \mathbf{y(x + 1) - z(x + 1)}$
$= \mathbf{(x + 1)(y - z)}$

(f) $a^2 - ab - a + b = \mathbf{a(a - b) - 1(a - b)}$
$= \mathbf{(a - b)(a - 1)}$

☞ $a(a - b) - (a - b)$ is the same thing.
Don't forget the 1 in this line.

(g) $k^2 - kt + 2k - 2t = \mathbf{k(k - t) + 2(k - t)}$
$= \mathbf{(k - t)(k + 2)}$

(h) $3x^2 + 9x + xy + 3y = \mathbf{3x(x + 3) + y(x + 3)}$
$= \mathbf{(x + 3)(3x + y)}$

(i) $ab - ac - bd + cd = \mathbf{a(b - c) - d(b - c)}$
$= \mathbf{(b - c)(a - d)}$

☞ Check that the signs are correct.

(j) $6t^2 + 2tu - 3t - u = \mathbf{2t(3t + u) - 1(3t + u)}$
$= \mathbf{(3t + u)(2t - 1)}$

(k) $abc - abd - c + d = \mathbf{ab(c - d) - 1(c - d)}$
$= \mathbf{(c - d)(ab - 1)}$

94 DIFFERENCE OF TWO SQUARES

When we multiply the sum of two terms by their difference we get a special result:

$(a + b)(a - b) = a^2 - b^2$ The result is the *difference of two squares.*

When factorised: $a^2 - b^2 = (a + b)(a - b)$

Example:

Factorise:

(a) $x^2 - 49$ **(b)** $a^2 - 64$ **(c)** $e^2 - 121$ **(d)** $h^2 - 1$ **(e)** $9 - b^2$ **(f)** $m^2 - n^2$

Solution:

(a) $x^2 - 49 = x^2 - 7^2$
$= (x + 7)(x - 7)$

(b) $a^2 - 64 = a^2 - 8^2$
$= (a + 8)(a - 8)$

(c) $e^2 - 121 = e^2 - 11^2$
$= (e + 11)(e - 11)$

(d) $h^2 - 1 = h^2 - 1^2$
$= (h + 1)(h - 1)$

(e) $9 - b^2 = 3^2 - b^2$
$= (3 + b)(3 - b)$

(f) $m^2 - n^2 = (m + n)(m - n)$

A square might be the product of two or more terms.

Example:

Factorise:

(a) $4x^2 - 25$ **(b)** $100 - 81z^2$ **(c)** $16x^2 - 49y^2$ **(d)** $x^4 - a^2b^2$

Solution:

(a) $4x^2 - 25 = (2x)^2 - 5^2$
$= (2x + 5)(2x - 5)$

(b) $100 - 81z^2 = 10^2 - (9z)^2$
$= (10 + 9z)(10 - 9z)$

(c) $16x^2 - 49y^2 = (4x)^2 - (7y)^2$
$= (4x + 7y)(4x - 7y)$

(d) $x^4 - a^2b^2 = (x^2)^2 - (ab)^2$
$= (x^2 + ab)(x^2 - ab)$

We might need to find a common factor first!

Example:

Factorise fully:

(a) $2x^2 - 72$ **(b)** $3a^2 - 27$ **(c)** $5d^2 - 20e^2$ **(d)** $16h^2 - 4$

Solution:

(a) $2x^2 - 72 = 2(x^2 - 36)$
$= 2(x^2 - 6^2)$
$= 2(x + 6)(x - 6)$

(b) $3a^2 - 27 = 3(a^2 - 9)$
$= 3(a^2 - 3^2)$
$= 3(a + 3)(a - 3)$

(c) $5d^2 - 20e^2 = 5(d^2 - 4e^2)$
$= 5(d^2 - (2e)^2)$
$= 5(d + 2e)(d - 2e)$

(d) $16h^2 - 4 = 4(4h^2 - 1)$
$= 4((2h)^2 - 1^2)$
$= 4(2h + 1)(2h - 1)$

1. **Complete:**

(a) $49 = \_\_\_^2$

(b) $144 = \_\_\_^2$

(c) $4x^2 = (\_\_\_)^2$

(d) $64p^2 = (\_\_\_)^2$

(e) $36t^2 = (\_\_\_)^2$

(f) $x^4 = (\_\_\_)^2$

(g) $16a^2b^2 = (\_\_\_\_)^2$

(h) $9t^4 = (\_\_\_)^2$

2. **Factorise:**

(a) $a^2 - b^2 =$ ________________

(b) $p^2 - 4^2 =$ ________________

(c) $k^2 - y^2 =$ ________________

(d) $e^2 - 1^2 =$ ________________

3. **Factorise:**

(a) $x^2 - 25 =$ ________________
= ________________

(b) $a^2 - 100 =$ ________________
= ________________

(c) $e^2 - 144 =$ ________________
= ________________

(d) $g^2 - 4 =$ ________________
= ________________

(e) $m^2 - 36 =$ ________________
= ________________

(f) $49 - b^2 =$ ________________
= ________________

(g) $k^2 - 1 =$ ________________
= ________________

(h) $p^2 - 81 =$ ________________
= ________________

(i) $16 - t^2 =$ ________________
= ________________

(j) $121 - y^2 =$ ________________
= ________________

(k) $4n^2 - 9 =$ ________________
= ________________

(l) $p^2 - 9q^2 =$ ________________
= ________________

(m) $64c^2 - 25d^2 =$ ________________
= ________________

(n) $1 - 49z^2 =$ ________________
= ________________

(o) $x^4 - y^2 =$ ________________
= ________________

(p) $a^4 - 4b^2 =$ ________________
= ________________

(q) $x^2y^2 - z^4 =$ ________________
= ________________

(r) $4 - 9p^2q^2 =$ ________________
= ________________

4. **Factorise, by first finding a common factor:**

(a) $2a^2 - 8 =$ ________________
= ________________
= ________________

(b) $5t^2 - 180 =$ ________________
= ________________
= ________________

(c) $7c^2 - 7d^2 =$ ________________
= ________________

(d) $3 - 3n^2 =$ ________________
= ________________

DIFFERENCE OF TWO SQUARES Worked Solutions

1. Complete:

(a) $49 = \mathbf{7^2}$

(b) $144 = \mathbf{12^2}$

(c) $4x^2 = (\mathbf{2x})^2$

(d) $64p^2 = (\mathbf{8p})^2$

(e) $36t^2 = (\mathbf{6t})^2$

(f) $x^4 = (\mathbf{x^2})^2$

(g) $16a^2b^2 = (\mathbf{4ab})^2$

(h) $9t^4 = (\mathbf{3t^2})^2$

2. Factorise:

(a) $a^2 - b^2 = \mathbf{(a + b)(a - b)}$ ☞ We have the difference of two squares.

(b) $p^2 - 4^2 = \mathbf{(p + 4)(p - 4)}$ ☞ We could put $(p - 4)(p + 4)$. It is the same thing.

(c) $k^2 - y^2 = \mathbf{(k + y)(k - y)}$

(d) $e^2 - 1^2 = \mathbf{(e + 1)(e - 1)}$

3. Factorise:

(a) $x^2 - 25 = \mathbf{x^2 - 5^2}$ ☞ First express as two squares.
$= \mathbf{(x + 5)(x - 5)}$

(b) $a^2 - 100 = \mathbf{a^2 - 10^2}$
$= \mathbf{(a + 10)(a - 10)}$

(c) $e^2 - 144 = \mathbf{e^2 - 12^2}$
$= \mathbf{(e + 12)(e - 12)}$

(d) $g^2 - 4 = \mathbf{g^2 - 2^2}$
$= \mathbf{(g + 2)(g - 2)}$

(e) $m^2 - 36 = \mathbf{m^2 - 6^2}$
$= \mathbf{(m + 6)(m - 6)}$

(f) $49 - b^2 = \mathbf{7^2 - b^2}$ ☞ Because the first term is 49, the first term in the factorisation must be 7. $(b + 7)(b - 7)$ is not the same thing and is incorrect.
$= \mathbf{(7 + b)(7 - b)}$

(g) $k^2 - 1 = \mathbf{k^2 - 1^2}$
$= \mathbf{(k + 1)(k - 1)}$

(h) $p^2 - 81 = \mathbf{p^2 - 9^2}$ ☞ You might not need to show this step.
$= \mathbf{(p + 9)(p - 9)}$

(i) $16 - t^2 = \mathbf{4^2 - t^2}$
$= \mathbf{(4 + t)(4 - t)}$

(j) $121 - y^2 = \mathbf{11^2 - y^2}$
$= \mathbf{(11 + y)(11 - y)}$

(k) $4n^2 - 9 = \mathbf{(2n)^2 - 3^2}$
$= \mathbf{(2n + 3)(2n - 3)}$

☞ $4n^2 = \mathbf{2^2 \times n^2 = (2n)^2}$

(l) $p^2 - 9q^2 = \mathbf{p^2 - (3q)^2}$
$= \mathbf{(p + 3q)(p - 3q)}$

☞ $9q^2 = \mathbf{3^2 \times q^2 = (3q)^2}$

(m) $64c^2 - 25d^2 = \mathbf{(8c)^2 - (5d)^2}$
$= \mathbf{(8c + 5d)(8c - 5d)}$

☞ $64c^2 = \mathbf{8^2 \times c^2 = (8c)^2}$ $25d^2 = \mathbf{5^2 \times d^2 = (5d)^2}$

(n) $1 - 49z^2 = \mathbf{1^2 - (7z)^2}$
$= \mathbf{(1 + 7z)(1 - 7z)}$

(o) $x^4 - y^2 = \mathbf{(x^2)^2 - y^2}$
$= \mathbf{(x^2 + y)(x^2 - y)}$

☞ Remember that $x^2 \times x^2 = \mathbf{x^4}$.

(p) $a^4 - 4b^2 = \mathbf{(a^2)^2 - (2b)^2}$
$= \mathbf{(a^2 + 2b)(a^2 - 2b)}$

(q) $x^2y^2 - z^4 = \mathbf{(xy)^2 - (z^2)^2}$
$= \mathbf{(xy + z^2)(xy - z^2)}$

☞ $(xy)^2 = \mathbf{xy \times xy = x^2y^2}$

(r) $4 - 9p^2q^2 = \mathbf{2^2 - (3pq)^2}$
$= \mathbf{(2 + 3pq)(2 - 3pq)}$

4. Factorise, by first finding a common factor:

(a) $2a^2 - 8 = \mathbf{2(a^2 - 4)}$
$= \mathbf{2(a^2 - 2^2)}$
$= \mathbf{2(a + 2)(a - 2)}$

☞ After taking out the common factor, 2, the part inside the bracket is the difference of two squares.

(b) $5t^2 - 180 = \mathbf{5(t^2 - 36)}$
$= \mathbf{5(t^2 - 6^2)}$
$= \mathbf{5(t + 6)(t - 6)}$

☞ $5t^2 - 180 = \mathbf{5 \times t^2 - 5 \times 36}$

Don't forget to write the 5 again.

(c) $7c^2 - 7d^2 = \mathbf{7(c^2 - d^2)}$
$= \mathbf{7(c + d)(c - d)}$

☞ Don't forget the 7.

(d) $3 - 3n^2 = \mathbf{3(1 - n^2)}$
$= \mathbf{3(1^2 - n^2)}$
$= \mathbf{3(1 + n)(1 - n)}$

98 FACTORISING TRINOMIALS

A trinomial has three terms.

We know that $(x + 5)(x + 3) = \mathbf{x^2 + 3x + 5x + 15}$
$= \mathbf{x^2 + 8x + 15}$

Factorising is the opposite of expanding,
so when factorised: $x^2 + 8x + 15 = \mathbf{(x + 5)(x + 3)}$.

Notice that the two numbers in the answer (5 and 3) add to 8 and multiply to 15.

When factorising a trinomial we need to look for two numbers that add to give the middle number and multiply to give the end number.

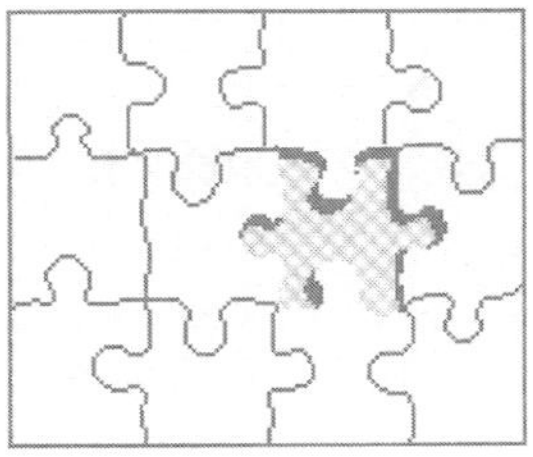

Each question is like a puzzle to solve!

Example:

Factorise:

(a) $x^2 + 10x + 24$ **(b)** $x^2 + 9x + 14$ **(c)** $x^2 + 6x + 5$

Solution:

(a) $x^2 + 10x + 24 = \mathbf{(x + 6)(x + 4)}$ — What 2 numbers add to 10 and multiply to 24?
(b) $x^2 + 9x + 14 = \mathbf{(x + 7)(x + 2)}$ — What 2 numbers add to 9 and multiply to 14?
(c) $x^2 + 6x + 5 = \mathbf{(x + 5)(x + 1)}$ — What numbers add to 6 and multiply to 5?

Sometimes the numbers that we are looking for might add or multiply to a negative number.

Example:

Factorise:

(a) $x^2 - 7x + 12$ **(b)** $x^2 - 11x + 30$ **(c)** $x^2 + 2x - 35$ **(d)** $x^2 - 3x - 70$

Solution:

(a) $x^2 - 7x + 12 = \mathbf{(x - 3)(x - 4)}$ — What 2 numbers add to –7 and multiply to 12?
(b) $x^2 - 11x + 30 = \mathbf{(x - 5)(x - 6)}$ — What numbers add to –11 and multiply to 30?
(c) $x^2 + 2x - 35 = \mathbf{(x + 7)(x - 5)}$ — What numbers add to 2 and multiply to –35?
(d) $x^2 - 3x - 70 = \mathbf{(x + 7)(x - 10)}$ — What numbers add to –3 and multiply to –70?

The order in which we write the answer doesn't matter as long as the signs are correct.

$x^2 + 7x - 60 = (x + 12)(x - 5)$ or $(x - 5)(x + 12)$

Example:

Factorise:

(a) $x^2 - x - 72$ **(b)** $x^2 + 8x + 16$ **(c)** $x^2 - 3x + 2$ **(d)** $x^2 + x - 12$

Solution:

(a) $x^2 - x - 72 = \mathbf{(x - 9)(x + 8)}$ — What numbers add to –1 and multiply to –72?
(b) $x^2 + 8x + 16 = \mathbf{(x + 4)(x + 4)}$ — We could write this as $(x + 4)^2$.
(c) $x^2 - 3x + 2 = \mathbf{(x - 2)(x - 1)}$
(d) $x^2 + x - 12 = \mathbf{(x + 4)(x - 3)}$

Not all trinomials can be factorised.
If you can't find an answer, the trinomial might not factorise.

1. **Complete:**

(a) $x^2 + 9x + 14 = (x + 7)(x + \_\_\_\_)$

(b) $x^2 + 11x + 30 = (x + \_\_\_\_)(x + 5)$

(c) $a^2 + 8a + 12 = (a + \_\_\_\_)(a + 6)$

(d) $k^2 + 15k + 56 = (k + 8)(k + \_\_\_\_)$

(e) $x^2 + 4x + 3 = (x + 3)(x + \_\_\_\_)$

(f) $y^2 + 7y + 12 = (y + \_\_\_\_)(y + \_\_\_\_)$

2. **Factorise:**

(a) $x^2 + 7x + 10 = \_\_\_\_\_\_\_\_\_\_\_\_$

(b) $x^2 + 10x + 21 = \_\_\_\_\_\_\_\_\_\_\_\_$

(c) $m^2 + 6m + 8 = \_\_\_\_\_\_\_\_\_\_\_\_$

(d) $a^2 + 5a + 6 = \_\_\_\_\_\_\_\_\_\_\_\_$

(e) $h^2 + 12h + 35 = \_\_\_\_\_\_\_\_\_\_\_\_$

(f) $n^2 + 20n + 96 = \_\_\_\_\_\_\_\_\_\_\_\_$

(g) $x^2 + 16x + 60 = \_\_\_\_\_\_\_\_\_\_\_\_$

(h) $p^2 + 6p + 5 = \_\_\_\_\_\_\_\_\_\_\_\_$

3. **Complete:**

(a) $x^2 - 10x + 16 = (x - 8)(x - \_\_\_\_)$

(b) $e^2 - 9e + 18 = (e - 6)(e - \_\_\_\_)$

(c) $x^2 + 3x - 10 = (x + 5)(x - \_\_\_\_)$

(d) $x^2 - 7x - 18 = (x + 2)(x - \_\_\_\_)$

(e) $q^2 - 13q + 40 = (q - \_\_\_\_)(q - 5)$

(f) $d^2 + d - 30 = (d + \_\_\_\_)(d - 5)$

(g) $x^2 - 4x - 21 = (x - 7)(x + \_\_\_\_)$

(h) $z^2 - 9z + 20 = (z - \_\_\_\_)(z - \_\_\_\_)$

4. **Factorise:**

(a) $x^2 - 10x + 24 = \_\_\_\_\_\_\_\_\_\_\_\_$

(b) $t^2 - 11t + 28 = \_\_\_\_\_\_\_\_\_\_\_\_$

(c) $x^2 - 2x - 15 = \_\_\_\_\_\_\_\_\_\_\_\_$

(d) $x^2 + 3x - 40 = \_\_\_\_\_\_\_\_\_\_\_\_$

(e) $s^2 - 8s + 15 = \_\_\_\_\_\_\_\_\_\_\_\_$

(f) $n^2 - n - 12 = \_\_\_\_\_\_\_\_\_\_\_\_$

(g) $a^2 - 6a - 27 = \_\_\_\_\_\_\_\_\_\_\_\_$

(h) $h^2 - 5h - 14 = \_\_\_\_\_\_\_\_\_\_\_\_$

(i) $b^2 - 12b + 32 = \_\_\_\_\_\_\_\_\_\_\_\_$

(j) $g^2 + 4g - 12 = \_\_\_\_\_\_\_\_\_\_\_\_$

5. **Factorise:**

(a) $x^2 + 9x + 8 = \_\_\_\_\_\_\_\_\_\_\_\_$

(b) $k^2 + 8k - 20 = \_\_\_\_\_\_\_\_\_\_\_\_$

(c) $t^2 - 4t - 45 = \_\_\_\_\_\_\_\_\_\_\_\_$

(d) $a^2 - 11a + 18 = \_\_\_\_\_\_\_\_\_\_\_\_$

(e) $m^2 - 14m + 49 = \_\_\_\_\_\_\_\_\_\_\_\_$

(f) $c^2 + 10c + 25 = \_\_\_\_\_\_\_\_\_\_\_\_$

(g) $e^2 + 18e - 40 = \_\_\_\_\_\_\_\_\_\_\_\_$

(h) $q^2 - 10q - 11 = \_\_\_\_\_\_\_\_\_\_\_\_$

FACTORISING TRINOMIALS Worked Solutions

1. Complete:

(a) $x^2 + 9x + 14 = (x + 7)(x + \mathbf{2})$ — What 2 numbers add to 9 and multiply to 14?

(b) $x^2 + 11x + 30 = (x + \mathbf{6})(x + 5)$ — What 2 numbers add to 11 and multiply to 30?

(c) $a^2 + 8a + 12 = (a + \mathbf{2})(a + 6)$ — What 2 numbers add to 8 and multiply to 12?

(d) $k^2 + 15k + 56 = (k + 8)(k + \mathbf{7})$ — What 2 numbers add to 15 and multiply to 56?

(e) $x^2 + 4x + 3 = (x + 3)(x + \mathbf{1})$ — What 2 numbers add to 4 and multiply to 3?

(f) $y^2 + 7y + 12 = (y + \mathbf{3})(y + \mathbf{4})$ — What 2 numbers add to 7 and multiply to 12? $(y + 4)(y + 3)$ is the same thing.

2. Factorise:

(a) $x^2 + 7x + 10 = \mathbf{(x + 5)(x + 2)}$ — What 2 numbers add to 7 and multiply to 10? $(x + 2)(x + 5)$ is the same thing.

(b) $x^2 + 10x + 21 = \mathbf{(x + 7)(x + 3)}$ — or $(x + 3)(x + 7)$

(c) $m^2 + 6m + 8 = \mathbf{(m + 2)(m + 4)}$ — What 2 numbers add to 6 and multiply to 8?

(d) $a^2 + 5a + 6 = \mathbf{(a + 3)(a + 2)}$

(e) $h^2 + 12h + 35 = \mathbf{(h + 7)(h + 5)}$ — What 2 numbers add to 12 and multiply to 35?

(f) $n^2 + 20n + 96 = \mathbf{(n + 12)(n + 8)}$

(g) $x^2 + 16x + 60 = \mathbf{(x + 10)(x + 6)}$ — or $(x + 6)(x + 10)$

(h) $p^2 + 6p + 5 = \mathbf{(p + 5)(p + 1)}$ — The numbers add to 6 and multiply to 5. Be careful with this question. $(p + 3)(p + 2)$ is not correct.

3. Complete:

(a) $x^2 - 10x + 16 = (x - 8)(x - \mathbf{2})$ — What 2 numbers add to –10 and multiply to 16?

(b) $e^2 - 9e + 18 = (e - 6)(e - \mathbf{3})$ — What 2 numbers add to –9 and multiply to 18?

(c) $x^2 + 3x - 10 = (x + 5)(x - \mathbf{2})$ — What 2 numbers add to 3 and multiply to –10?

(d) $x^2 - 7x - 18 = (x + 2)(x - \mathbf{9})$ — What 2 numbers add to –7 and multiply to –18?

(e) $q^2 - 13q + 40 = (q - \mathbf{8})(q - 5)$ — What 2 numbers add to –13 and multiply to 40?

(f) $d^2 + d - 30 = (d + 6)(d - 5)$ — What 2 numbers add to 1 and multiply to –30?

(g) $x^2 - 4x - 21 = (x - 7)(x + 3)$ — What 2 numbers add to –4 and multiply to –21?

(h) $z^2 - 9z + 20 = (z - 5)(z - 4)$ — $(z - 4)(z - 5)$ is the same thing.

4. Factorise:

(a) $x^2 - 10x + 24 = (x - 6)(x - 4)$ — What 2 numbers add to –10 and multiply to 24?

(b) $t^2 - 11t + 28 = (t - 7)(t - 4)$ — What 2 numbers add to –11 and multiply to 28?

(c) $x^2 - 2x - 15 = (x - 5)(x + 3)$ — What 2 numbers add to –2 and multiply to –15?

(d) $x^2 + 3x - 40 = (x + 8)(x - 5)$ — What 2 numbers add to 3 and multiply to –40?

(e) $s^2 - 8s + 15 = (s - 5)(s - 3)$ — What 2 numbers add to –8 and multiply to 15?

(f) $n^2 - n - 12 = (n - 4)(n + 3)$ — What 2 numbers add to –1 and multiply to –12?

(g) $a^2 - 6a - 27 = (a - 9)(a + 3)$ — $(a + 3)(a - 9)$ is the same thing.

(h) $h^2 - 5h - 14 = (h - 7)(h + 2)$ — What 2 numbers add to –5 and multiply to –14?

(i) $b^2 - 12b + 32 = (b - 8)(b - 4)$ — or $(b - 4)(b - 8)$

(j) $g^2 + 4g - 12 = (g + 6)(g - 2)$

5. Factorise:

(a) $x^2 + 9x + 8 = (x + 8)(x + 1)$ — What 2 numbers add to 9 and multiply to 8?

(b) $k^2 + 8k - 20 = (k + 10)(k - 2)$

(c) $t^2 - 4t - 45 = (t - 9)(t + 5)$ — What 2 numbers add to –4 and multiply to –45?

(d) $a^2 - 11a + 18 = (a - 9)(a - 2)$

(e) $m^2 - 14m + 49 = (m - 7)(m - 7)$ — or $(m - 7)^2$

(f) $c^2 + 10c + 25 = (c + 5)(c + 5)$ — or $(c + 5)^2$

(g) $e^2 + 18e - 40 = (e + 20)(e - 2)$

(h) $q^2 - 10q - 11 = (q - 11)(q + 1)$

102 MIXED FACTORISING

We can now complete several different types of factorisation: common factors; factorising by grouping; difference of two squares; trinomials.

Things to consider: First, is there a common factor?
If there is no common factor, but there are:
4 terms, can they be factorised by grouping in pairs?
3 terms, is it a trinomial that can be factorised?
2 terms, are they the difference of two terms?

Example:

Factorise:
(a) $3x + 12$ **(b)** $x^2 + 7x + 10$ **(c)** $ab + 5a + 4b + 20$ **(d)** $x^2 - 64$ **(e)** $14k + 7m - 35$ **(f)** $x^2 - 25x$

Solution:

(a) $3x + 12 = \mathbf{3(x + 4)}$

(b) $x^2 + 7x + 10 = \mathbf{(x + 5)(x + 2)}$

(c) $ab + 5a + 4b + 20 = a(b + 5) + 4(b + 5)$
$= \mathbf{(b + 5)(a + 4)}$

(d) $x^2 - 64 = x^2 - 8^2$
$= \mathbf{(x + 8)(x - 8)}$

(e) $14k + 7m - 35 = \mathbf{7(2k + m - 5)}$

(f) $x^2 - 25x = \mathbf{x(x - 25)}$

We must always factorise as fully as possible.
This might require the combination of different types of factorisation.
Always look for a common factor first.

Example:

Factorise fully:
(a) $2x^2 - 8$ **(b)** $3x^2 - 12x + 9$ **(c)** $2x^2 + 8x + 2xy + 8y$ **(d)** $y - y^3$ **(e)** $x^2 - 16 + xy - 4y$ **(f)** $81 - t^4$

Solution:

(a) $2x^2 - 8 = 2(x^2 - 4)$
$= 2(x^2 - 2^2)$
$= \mathbf{2(x + 2)(x - 2)}$

(b) $3x^2 - 12x + 9 = 3(x^2 - 4x + 3)$
$= \mathbf{3(x - 3)(x - 1)}$

(c) $2x^2 + 8x + 2xy + 8y = 2(x^2 + 4x + xy + 4y)$
$= 2(x(x + 4) + y(x + 4))$
$= \mathbf{2(x + 4)(x + y)}$

(d) $y - y^3 = y(1 - y^2)$
$= \mathbf{y(1 + y)(1 - y)}$

(e) $x^2 - 16 + xy - 4y = x^2 - 4^2 + y(x - 4)$
$= (x + 4)(x - 4) + y(x - 4)$
$= \mathbf{(x - 4)(x + 4 + y)}$

(f) $81 - t^4 = 9^2 - (t^2)^2$
$= (9 + t^2)(9 - t^2)$
$= (9 + t^2)(3^2 - t^2)$
$= \mathbf{(9 + t^2)(3 + t)(3 - t)}$

1. **Factorise by finding a common factor:**

(a) $9k - 81 =$ ______
(c) $x^2 + 8x =$ ______
(e) $15t - 20u =$ ______
(b) $8 + 12m =$ ______
(d) $5p + 10q =$ ______
(f) $6a^2 + 9a =$ ______

2. **Factorise by grouping:**

(a) $x^2 + xy + 5x + 5y$
(b) $ab - bc + ac - c^2$
(c) $pq - p - q + 1$

3. **Factorise each difference of two squares:**

(a) $x^2 - 81 =$ ______ = ______
(c) $p^2 - 1 =$ ______ = ______
(e) $a^4 - 25b^2 =$ ______ = ______
(b) $64 - n^2 =$ ______ = ______
(d) $z^2 - 9 =$ ______ = ______
(f) $9x^2 - 16y^2 =$ ______ = ______

4. **Factorise each trinomial:**

(a) $e^2 - 9e - 22 =$ ______
(c) $m^2 + m - 72 =$ ______
(b) $h^2 + 13h + 42 =$ ______
(d) $x^2 - 6x + 9 =$ ______

5. **Factorise:**

(a) $9y - 6p$
(d) $x^2 + 8x - 33$
(g) $x^2 - 9$
(b) $d^2 - 36$
(e) $t^2 + 5t + tu + 5u$
(h) $x^2 - 9x + 14$
(c) $x^2 + xy + 2x$
(f) $x^2 - 9x$
(i) $x^2 - 9x + 9xy - 81y$

6. **Factorise:**

(a) $2a^2 - 32$
(c) $x^2 - 36 + xy + 6y$
(e) $28 - 63y^2$
(b) $5e^2 + 25e + 30$
(d) $3p^2 - 6pq + 3p - 6q$
(f) $p^4 - 16$

104 MIXED FACTORISING Worked Solutions

1. Factorise, by finding a common factor:

   (a) $9k - 81 = \mathbf{9(k - 9)}$ ☞ $9k - 81 = \mathbf{9 \times k - 9 \times 9}$

   (b) $8 + 12m = \mathbf{4(2 + 3m)}$ ☞ The largest common factor is 4.

   (c) $x^2 + 8x = \mathbf{x(x + 8)}$ ☞ $x^2 + 8x = \mathbf{x \times x + x \times 8}$

   (d) $5p + 10q = \mathbf{5(p + 2q)}$ (e) $15t - 20u = \mathbf{5(3t - 4u)}$ (f) $6a^2 + 9a = \mathbf{3a(2a + 3)}$

2. Factorise by grouping:

   (a) $x^2 + xy + 5x + 5y = \mathbf{x(x + y) + 5(x + y)}$ ☞ The first 2 terms have a common factor of x and the second 2 terms have a different common factor of 5.

   $= \mathbf{(x + y)(x + 5)}$ Taking out the common factor of $(x + y)$.

   (b) $ab - bc + ac - c^2 = \mathbf{b(a - c) + c(a - c)}$
   $= \mathbf{(a - c)(b + c)}$

   (c) $pq - p - q + 1 = \mathbf{p(q - 1) - 1(q - 1)}$ ☞ We need to be very careful with the signs.
   $= \mathbf{(q - 1)(p - 1)}$

3. Factorise each difference of two squares:

   (a) $x^2 - 81 = \mathbf{x^2 - 9^2}$ ☞ First express as two squares.
   $= \mathbf{(x + 9)(x - 9)}$

   (b) $64 - n^2 = \mathbf{8^2 - n^2}$
   $= \mathbf{(8 + n)(8 - n)}$

   (c) $p^2 - 1 = \mathbf{p^2 - 1^2}$
   $= \mathbf{(p + 1)(p - 1)}$

   (d) $z^2 - 9 = \mathbf{z^2 - 3^2}$
   $= \mathbf{(z + 3)(z - 3)}$

   (e) $a^4 - 25b^2 = \mathbf{(a^2)^2 - (5b)^2}$ ☞ $a^4 = \mathbf{(a^2)^2}$ $25b^2 = \mathbf{5^2 \times b^2 = (5b)^2}$
   $= \mathbf{(a^2 + 5b)(a^2 - 5b)}$

   (f) $9x^2 - 16y^2 = \mathbf{(3x)^2 - (4y)^2}$
   $= \mathbf{(3x + 4y)(3x - 4y)}$

4. Factorise each trinomial:

   (a) $e^2 - 9e - 22 = \mathbf{(e - 11)(e + 2)}$ ☞ What 2 numbers add to –9 and multiply to –22?

   (b) $h^2 + 13h + 42 = \mathbf{(h + 6)(h + 7)}$ ☞ What 2 numbers add to 13 and multiply to 42?

   (c) $m^2 + m - 72 = \mathbf{(m + 9)(m - 8)}$ ☞ What 2 numbers add to 1 and multiply to –72?

   (d) $x^2 - 6x + 9 = \mathbf{(x - 3)(x - 3)}$ ☞ or $(x - 3)^2$

5. Factorise:

(a) $9y - 6p = 3(3y - 2p)$ ☞ 3 is a common factor.

(b) $d^2 - 36 = d^2 - 6^2$ ☞ Difference of two squares.
$= (d + 6)(d - 6)$

(c) $x^2 + xy + 2x = x(x + y + 2)$ ☞ x is a common factor.

(d) $x^2 + 8x - 33 = (x - 3)(x + 11)$ ☞ What 2 numbers add to 8 and multiply to –33?

(e) $t^2 + 5t + tu + 5u = t(t + 5) + u(t + 5)$ ☞ Factorising by grouping.
$= (t + 5)(t + u)$

(f) $x^2 - 9x = x(x - 9)$ ☞ x is a common factor.

(g) $x^2 - 9 = x^2 - 3^2$ ☞ The difference of two squares.
$= (x + 3)(x - 3)$

(h) $x^2 - 9x + 14 = (x - 2)(x - 7)$ ☞ or $(x - 7)(x - 2)$

(i) $x^2 - 9x + 9xy - 81y = x(x - 9) + 9y(x - 9)$ ☞ Factorising by grouping.
$= (x - 9)(x + 9y)$

6. Factorise:

(a) $2a^2 - 32 = 2(a^2 - 16)$ ☞ First take out a common factor,
$= 2(a + 4)(a - 4)$ then we have the difference of two squares.

(b) $5e^2 + 25e + 30 = 5(e^2 + 5e + 6)$ ☞ First take out the common factor, 5,
$= 5(e + 3)(e + 2)$ then find 2 numbers that add to 5 and multiply to 6.

(c) $x^2 - 36 + xy + 6y = x^2 - 6^2 + xy + 6y$ ☞ Difference of two squares; common factor of y.
$= (x + 6)(x - 6) + y(x + 6)$
$= (x + 6)(x - 6 + y)$ Taking out a common factor of $(x + 6)$.

(d) $3p^2 - 6pq + 3p - 6q$
$= 3(p^2 - 2pq + p - 2q)$ ☞ First take out the common factor, 3,
$= 3[p(p - 2q) + 1(p - 2q)]$ then factorise by grouping.
$= 3(p - 2q)(p + 1)$

(e) $28 - 63y^2 = 7(4 - 9y^2)$ ☞ First take out the common factor, 7,
$= 7(2^2 - (3y)^2)$ then we have the difference of two squares.
$= 7(2 + 3y)(2 - 3y)$

(f) $p^4 - 16 = (p^2)^2 - 4^2$ ☞ We have the difference of two squares,
$= (p^2 + 4)(p^2 - 4)$
$= (p^2 + 4)(p^2 - 2^2)$ and another difference of two squares.
$= (p + 4)(p + 2)(p - 2)$

Fractions such as $\frac{6e + 18f}{6}$ can be simplified by cancelling, but extreme care must be taken to do this correctly.

$$\frac{6e + 18f}{6} = \frac{{}^{1}\cancel{6}(e + 3f)}{\cancel{6}_{1}} = e + 3f$$

Take out a common factor so as not to make a common mistake!

Example:

Factorise and simplify:

(a) $\frac{9a - 6ab}{3a}$ **(b)** $\frac{14xy - 21x}{7x}$ **(c)** $\frac{2x^2 - 6x}{8x}$

Solution:

(a) $\frac{9a - 6ab}{3a} = \frac{{}^{1}\cancel{3a}(3 - 2b)}{\cancel{3a}_{1}} = \mathbf{3 - 2b}$

(b) $\frac{14xy + 21x}{7x} = \frac{{}^{1}\cancel{7x}(2y + 3)}{\cancel{7x}_{1}} = \mathbf{2y + 3}$

(c) $\frac{2x^2 - 6x}{8x} = \frac{{}^{1}\cancel{2x}(x - 3)}{\cancel{8x}_{4}} = \mathbf{\frac{x - 3}{4}}$

We can leave the answer in factorised form.

Example:

Factorise and simplify:

(a) $\frac{3x}{5x + x^2}$ **(b)** $\frac{6}{36x - 9}$ **(c)** $\frac{2a}{4ab + 14b}$

Solution:

(a) $\frac{3x}{5x + x^2} = \frac{{}^{3}\cancel{3x}}{{}_{1}\cancel{x}(5 + x)} = \mathbf{\frac{3}{5 + x}}$

(b) $\frac{6}{36x - 9} = \frac{{}^{2}\cancel{6}}{{}_{3}\cancel{9}(4x - 1)} = \mathbf{\frac{2}{3(4x - 1)}}$

(c) $\frac{2a}{4ab + 14b} = \frac{{}^{1}\cancel{2}a}{{}_{1}\cancel{2}b(2a + 7)} = \mathbf{\frac{a}{b(2a + 7)}}$

Example:

Never leave out a '1' in the numerator.

Factorise and simplify:

(a) $\frac{m}{mn - 6m}$ **(b)** $\frac{5}{10x + 15y}$ **(c)** $\frac{e}{e^2 - ef}$ **(d)** $\frac{2}{12x + 8y}$

Solution:

(a) $\frac{m}{mn - 6m} = \frac{{}^{1}\cancel{m}}{{}_{1}\cancel{m}(n - 6)} = \mathbf{\frac{1}{n - 6}}$

(b) $\frac{5}{10x + 15y} = \frac{{}^{1}\cancel{5}}{{}_{1}\cancel{5}(2x + 3y)} = \mathbf{\frac{1}{2x + 3y}}$

(c) $\frac{e}{e^2 - ef} = \frac{{}^{1}\cancel{e}}{{}_{1}\cancel{e}(e - f)} = \mathbf{\frac{1}{e - f}}$

(d) $\frac{2}{12x + 8y} = \frac{{}^{1}\cancel{2}}{{}_{2}\cancel{4}(3x + 2y)} = \mathbf{\frac{1}{2(3x + 2y)}}$

Example:

Factorise and simplify:

(a) $\frac{x^2 - 9x}{x^2 - 81}$ **(b)** $\frac{x^2 + 8x + 12}{x^2 + 4x - 12}$

We can't cancel any term that is not a common factor. $[\frac{x + 3}{x} \neq 3]$

Solution:

(a) $\frac{x^2 - 9x}{x^2 - 81} = \frac{x\cancel{(x - 9)}^{1}}{(x + 9)\cancel{(x - 9)}_{1}} = \mathbf{\frac{x}{x + 9}}$

(b) $\frac{x^2 + 8x + 12}{x^2 + 4x - 12} = \frac{{}^{1}\cancel{(x + 6)}(x + 2)}{{}_{1}\cancel{(x + 6)}(x - 2)} = \mathbf{\frac{x + 2}{x - 2}}$

1. **Simplify, by cancelling:**

(a) $\frac{4(2a+b)}{2}$ = ____________

(b) $\frac{9(x+2)}{27}$ = ____________

(c) $\frac{x(x-4)}{x}$ = ____________

(d) $\frac{6t(t-u)}{2t}$ = ____________

(e) $\frac{8y(y-7)}{y-7}$ = ____________

(f) $\frac{3xy(x+y)}{9x^2}$ = ____________

(g) $\frac{e-f}{3(e-f)}$ = ____________

(h) $\frac{4}{8(x-4)}$ = ____________

(i) $\frac{(x+3)(x+2)}{(x-3)(x+3)}$ = ____________

2. **Factorise and simplify:**

(a) $\frac{8x+12}{4}$ = ____________
= ____________

(b) $\frac{5t-10u}{15}$ = ____________
= ____________

(c) $\frac{d^2-7d}{7d}$ = ____________
= ____________

(d) $\frac{36}{6p-12q}$ = ____________
= ____________

(e) $\frac{9a}{a^2-ab}$ = ____________
= ____________

(f) $\frac{-5}{20m+25n}$ = ____________
= ____________

(g) $\frac{2}{6x-8y}$ = ____________
= ____________

(h) $\frac{7y-14}{7}$ = ____________
= ____________

(i) $\frac{12}{9k-12}$ = ____________
= ____________

(j) $\frac{d}{ad+bd}$ = ____________
= ____________

(k) $\frac{3h}{h^2-9h}$ = ____________
= ____________

(l) $\frac{b^2-5b}{5b}$ = ____________
= ____________

(m) $\frac{3m+3n}{5m+5n}$ = ____________
= ____________

(n) $\frac{p^2-8p}{3p-24}$ = ____________
= ____________

(o) $\frac{6x-9}{2xy-3y}$ = ____________
= ____________

(p) $\frac{x^2-36}{2x-12}$ = ____________
= ____________

(q) $\frac{x^2+17x+70}{x^2+7x-30}$ = ____________
= ____________

(r) $\frac{x^2-4x}{x^2-16}$ = ____________
= ____________

(s) $\frac{x^2-25}{x^2-10x+25}$ = ____________
= ____________

1. Simplify, by cancelling:

(a) $\frac{4(2a+b)}{2} = \frac{^{2}\cancel{4}(2a+b)}{_{1}\cancel{2}}$

$= \mathbf{2(2a+b)}$

2 divides into both 4 and 2.

(b) $\frac{9(x+2)}{27} = \frac{^{1}\cancel{9}(x+2)}{_{3}\cancel{27}}$

$= \frac{\mathbf{x+2}}{\mathbf{3}}$

9 divides into both 9 and 27.

(c) $\frac{x(x-4)}{x} = \frac{^{1}\cancel{x}(x-4)}{_{1}\cancel{x}}$

$= \mathbf{x-4}$

x divides into both numerator and denominator.

(d) $\frac{6t(t-u)}{2t} = \frac{^{3}\cancel{6t}(t-u)}{_{1}\cancel{2t}}$

$= \mathbf{3(t-u)}$

Both 3 and *t* divide into the numerator and denominator.

(e) $\frac{8y(y-7)}{y-7} = \frac{8y\cancel{(y-7)}^{1}}{_{1}\cancel{y-7}}$

$= \mathbf{8y}$

$(y - 7)$ divides into both numerator and denominator.

(f) $\frac{3xy(x+y)}{9x^2} = \frac{^{1}\cancel{3x}y(x+y)}{_{3x}\cancel{9x^2}}$

$= \frac{\mathbf{y(x+y)}}{\mathbf{3x}}$

3*x* divides into both numerator and denominator.

(g) $\frac{e-f}{3(e-f)} = \frac{^{1}\cancel{e-f}}{3\cancel{(e-f)}_{1}}$

$= \frac{\mathbf{1}}{\mathbf{3}}$

$(e - f)$ divides into both numerator and denominator.

Don't forget the 1 in the numerator.

(h) $\frac{4}{8(x-4)} = \frac{\cancel{4}^{1}}{_{2}\cancel{8}(x-4)}$

$= \frac{\mathbf{1}}{\mathbf{(2x-4)}}$

4 divides into both numerator and denominator.

Don't forget the 1 in the numerator.

(i) $\frac{(x+3)(x+2)}{(x-3)(x+3)} = \frac{^{1}\cancel{(x+3)}(x+2)}{(x-3)\cancel{(x+3)}_{1}}$

$= \frac{\mathbf{x+2}}{\mathbf{x-3}}$

$(x + 3)$ is in both numerator and denominator.

The *x*s cannot cancel. They are not common factors.

2. Factorise and simplify:

(a) $\frac{8x+12}{4} = \frac{^{1}\cancel{4}(2x+3)}{\cancel{4}_{1}}$

$= \mathbf{2x+3}$

4 is a common factor. We can cancel out the 4s.

(b) $\frac{5t-10u}{15} = \frac{^{1}\cancel{5}(t-2u)}{\cancel{15}_{3}}$

$= \frac{\mathbf{t-2u}}{\mathbf{3}}$

5 is a common factor.

5 divides into both numerator and denominator.

(c) $\frac{d^2-7d}{7d} = \frac{^{1}\cancel{d}(d-7)}{7\cancel{d}_{1}}$

$= \frac{\mathbf{d-7}}{\mathbf{7}}$

d divides into both numerator and denominator.

The 7s can't be cancelled. 7 is not a common factor.

(d) $\frac{36}{6p-12q} = \frac{^{6}\cancel{36}}{_{1}\cancel{6}(p-2q)}$

$= \frac{\mathbf{6}}{\mathbf{p-2q}}$

(e) $\dfrac{9a}{a^2 - ab} = \dfrac{9\cancel{a}^{1}}{{}_{1}\cancel{a}(a-b)} = \dfrac{9}{a-b}$

(f) $\dfrac{-5}{20m + 25n} = \dfrac{-\cancel{5}^{-1}}{{}_{1}\cancel{5}(4m+5n)} = \dfrac{-1}{4m+5n}$

(g) $\dfrac{2}{6x - 8y} = \dfrac{\cancel{2}^{1}}{{}_{1}\cancel{2}(3x-4y)} = \dfrac{1}{3x-4y}$

(h) $\dfrac{7y - 14}{7} = \dfrac{{}^{1}\cancel{7}(y-2)}{\cancel{7}_{1}} = y - 2$

(i) $\dfrac{12}{9k - 12} = \dfrac{{}^{4}\cancel{12}}{{}_{1}\cancel{3}(3k-4)} = \dfrac{4}{3k-4}$

(j) $\dfrac{d}{ad + bd} = \dfrac{{}^{1}\cancel{d}}{{}_{1}\cancel{d}(a+b)} = \dfrac{1}{a+b}$

(k) $\dfrac{3h}{h^2 - 9h} = \dfrac{3\cancel{h}^{1}}{{}_{1}\cancel{h}(h-9)} = \dfrac{3}{h-9}$

(l) $\dfrac{b^2 - 5b}{5b} = \dfrac{{}^{1}\cancel{b}(b-5)}{5\cancel{b}_{1}} = \dfrac{b-5}{5}$

(m) $\dfrac{3m + 3n}{5m + 5n} = \dfrac{3\cancel{(m+n)}^{1}}{5\cancel{(m+n)}_{1}} = \dfrac{3}{5}$

3 is a common factor in the numerator and
5 is a common factor in the denominator.
$(m + n)$ divides into both numerator and denominator.

(n) $\dfrac{p^2 - 8p}{3p - 24} = \dfrac{p\cancel{(p-8)}^{1}}{3\cancel{(p-8)}_{1}} = \dfrac{p}{3}$

(o) $\dfrac{6x - 9}{2xy - 3y} = \dfrac{3\cancel{(2x-3)}^{1}}{y\cancel{(2x-3)}_{1}} = \dfrac{3}{y}$

(p) $\dfrac{x^2 - 36}{2x - 12} = \dfrac{(x+6)\cancel{(x-6)}^{1}}{2\cancel{(x-6)}_{1}} = \dfrac{x+6}{2}$

The numerator is the difference of two squares.
The denominator has a common factor of 2.

This cannot be simplified. (2 is not a common factor.)

(q) $\dfrac{x^2 + 17x + 70}{x^2 + 7x - 30} = \dfrac{{}^{1}\cancel{(x+10)}(x+7)}{{}_{1}\cancel{(x+10)}(x-3)} = \dfrac{x+7}{x-3}$

Both numerator and denominator are trinomials.

The *x*s cannot cancel. They are not common factors.

(r) $\dfrac{x^2 - 4x}{x^2 - 16} = \dfrac{x\cancel{(x-4)}^{1}}{(x+4)\cancel{(x-4)}_{1}} = \dfrac{x}{x+4}$

(s) $\dfrac{x^2 - 25}{x^2 - 10x + 25} = \dfrac{(x+5)\cancel{(x-5)}^{1}}{(x-5)\cancel{(x-5)}_{1}} = \dfrac{x+5}{x-5}$

Algebra	A branch of maths that uses symbols.
Algebraic expression	Any term, or combination of terms, using pronumerals.
Base	(In indices) the number, or term, that is multiplied. (5^2; 5 is the base.)
Binomial	An expression which is the sum, or difference, of two terms.
Binomial product	The multiplication of binomials.
Collecting like terms	Adding or subtracting terms.
Common denominator	Of fractions, a number divisible by all the denominators. (12 is the lowest common denominator of $\frac{1}{4}$ and $\frac{5}{6}$).
Common factors	Divide evenly into all the given numbers (e.g. 3 is a common factor of 9 and 12).
Consecutive numbers	Numbers that follow one another in order, (like 11, 12, 13, 14).
Denominator	The bottom part of a fraction. (For the fraction $\frac{2}{3}$, 3 is the denominator.)
Difference of two squares	A special binomial product. One square minus another square.
Distributive law	The law that says, for example, $2 \times (4 + 7) = 2 \times 4 + 2 \times 7$
Equation	An algebraic expression that includes an "=" sign.
Expand	To write out in longer form.
Factor	A number (or term) that divides into another. (7 is a factor of 21.)
Factorising	Writing something as a product of its factors (the opposite of expanding).
Formula	A rule using pronumerals (e.g. $A = \pi r^2$ is the formula for the area of a circle).
Function (of *x*)	A rule using a pronumeral (x). (Replacing x with a number gives a value of the function.)
$f(x)$	A function of x. We say "f of x".
Grouping symbols	Commonly called brackets: (parentheses), [brackets] and {braces}.
Highest common factor	The largest number that divides into all the given numbers.
Index	The number of times something is multiplied by itself. (3^4; 4 is the index.)
Index notation	Expressed with indices.
Indices	The plural of index (1 index, 2 or more indices).
Inequality	Something unequal (e.g. $6 > 4$, $x - 2 \leq -1$, $8 \neq 0$).
Inequation	An expression involving an inequality sign. (For example: $3x + 1 > 7$.)
Like terms	Terms that have exactly the same pronumeral part.
Number line	A line with integers (whole numbers) marked along it in order.
Numeral	A number.

Numerator	The top part of a fraction. (For the fraction $\frac{2}{3}$, 2 is the numerator.)
Operations	Addition, subtraction, multiplication and division.
Plotting	Placing points on a grid or a number line.
Power	Another term for an index.
Product	The result of a multiplication.
Pronumeral	A symbol that stands in place of a number.
Reciprocals	Multiply to give 1. ($\frac{3}{2}$ is the reciprocal of $\frac{2}{3}$.)
Satisfies	Makes a statement true. ($x = 2$ satisfies $x + 1 = 3$.)
Simplify	Make simpler, usually by adding, subtracting, etc.
Solution	The answer to an equation, or to a problem.
Square	A number, or term, multiplied by itself.
Subject	The pronumeral before the equals sign in a formula. ($E = mc^2$, E is the subject.)
Substitution	Replacing a pronumeral with a number.
Sum by difference	A binomial product where one binomial is the sum of 2 terms and the other is the difference of the same 2 terms.
Table of values	Values obeying a given rule, placed in a table.
Term	Each of the members of an algebraic expression (like $2x$, $8y^2$, 7, $9ab$).
Trinomial	An expression with three terms.
Unlike terms	Terms whose pronumeral parts are not exactly the same.

NOTES

NOTES

NOTES